Electrical Engineering Basics for HVAC System

FIRST EDITION

BY PRASUN BARUA

ABOUT

Welcome to Electrical Engineering Basics for HVAC System! This is a nonfiction science book which covers fundamentals of electrical engineering that HVAC engineers and other mechanical engineers would find beneficial in their daily work. A heating, ventilation, and air-conditioning (HVAC) system is a simple system of heating and cooling exchangers that use water or refrigerant as the medium (direct expansion system). Pumps transport warm or cooled water to exchangers. The warmed or cooled air produced by the exchangers is subsequently moved to the occupied building interiors via fans. As a result, heating and cooling occur in two stages. They are as follows: water stage and air stage. Water is the most efficient and cost-effective medium for cooling (through a chiller) or heating (via a boiler). Because it may be cooled or heated through coils, air is the medium for heat exchange in the building. Fans are utilized in HVAC systems for air circulation and ventilation. Cooling is performed through the use of chillers to provide chilled water for large buildings or direct expansion cooling devices such as packaged air conditioners for small structures. In the heating system, boilers are most commonly used to provide hot water for heating, but electric heaters are also widely utilized for zonal reheat. Pumps are used to circulate hot water, chilled water, and condenser water. Heat rejection is accomplished by the use of cooling towers. The cooling tower fan and pumps consume the most energy. Non-electrical engineers may find power distribution systems and equipment needed to power HVAC machines, motors, and other auxiliaries to be difficult. This is the first edition of the book. Thanks for reading the book.

TABLE OF CONTENTS

CHAPTER NO.	TITLE	PAGE NO.
CHAPTER-1	ELECTRICAL ENGINEERING BASICS	4
CHAPTER-2	POWER DISTRIBUTION ELEMENTS	25
CHAPTER-3	MOTORS & VARIABLE SPEED DRIVES	64
CHAPTER-4	EFFICIENCY OF ELECTRICAL ENERGY	88

CHAPTER-1: ELECTRICAL ENGINEERING BASICS

An electric prime mover (motor) drives the mechanical movement of various pieces of equipment in each facility. Electrical power is generated by utilities or internal generators and distributed via transformers to provide usable voltage levels. Electricity is commonly found in two forms:

1. AC (alternating current)
2. DC (direct current)

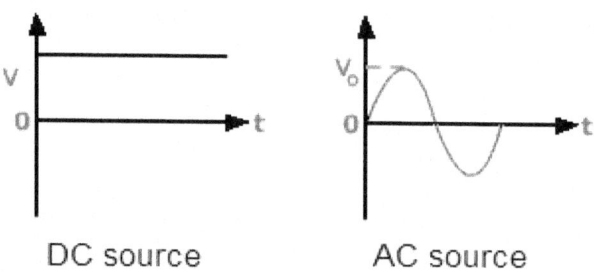

DC source   AC source

The choice of an energy source for equipment is determined by its application; each has advantages and disadvantages, but for an HVAC system or normal building services, we are concerned with AC voltage. Industrial alternating current voltage levels are typically classified as LV (low voltage) and HV (high voltage), with frequencies ranging from 50 to 60 Hz. Regardless of the kind of electrical energy, an electrical circuit contains three essential components:

1. Voltage (V) is defined as the electrical potential difference that causes electrons to flow.
2. Current (I) is defined as the flow of electrons and is measured in amperes.
3. Resistance (R) is defined as the opposition to the flow of electrons and is measured in ohms.

All three are bound together with Ohm's law, which gives the following relation between the three:

$$V = I \times R$$

In a more technical expression, you can state it as:

With Constant Resistance

- Lower voltage gives small current.
- Higher voltage gives large current.

With Constant Voltage

- Lower resistance passes large current.
- Higher resistance passes small current.
- **Example:** A conductor has a resistance of 1.5 ohms and the current flowing on the wire is 5 amperes. The

voltage drop along the wire is the current times the resistance of the conductor or 7.5 volts.

Example: A resistance type heating element from an electric water heater operating at 240 volts has a current flow of 14.6 amperes. The resistance of the heating element is the voltage divided by the current or 16.4 ohms.

CIRCUITS

In order to flow, electricity must have a continuous, closed path from start to finish; like a circle. The word "circuit" refers to the entire course an electric current travel, from the source of power, through an electrical device, and back to the source. Every circuit is comprised of three major components:

1. A conductive "path," such as a wire, or printed etches on a circuit board;
2. A "source" of electrical power, such as a battery or household wall outlet; and
3. A "load" that needs electrical power to operate, such as a lamp.

The current flows to the devices (called loads) via a "hot" wire and returns via a "neutral" wire since it is normally maintained at zero volts, also known as ground potential. In addition, two alternative components can be inserted in an electrical circuit. These are both control and defensive devices. However, control and safety mechanisms are not required for a circuit to function. They are entirely optional. A circuit that turns on an air conditioner when the temperature is too high, for example, would include the following components:

- a source of electrical energy, in this case, simple household current;
- a protective device that senses current flow on the circuit, the circuit breaker in the panel box;
- a control device that redirects the current, the switch in the thermostat; and
- a load such as an air conditioner that cools the space down until the circuit opens shutting the air conditioner off.

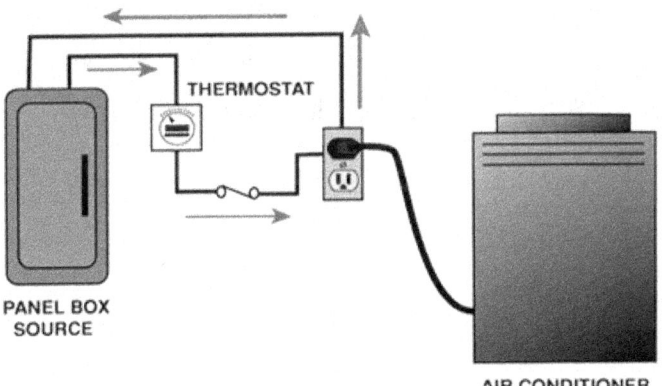

PANEL BOX
SOURCE

AIR CONDITIONER

Types of Circuits:

Circuits are classified into numerous categories. Their names describe the circuit's wiring or its primary function. The following are examples of basic circuits:

- Series
- Parallel
- Open
- Short
- Power
- Control

Series Circuit:

A series circuit is one in which the elements in a series carry the same current but have distinct voltage drops across them. The circuit has only one way for current to travel across it. A typical series circuit is illustrated below:

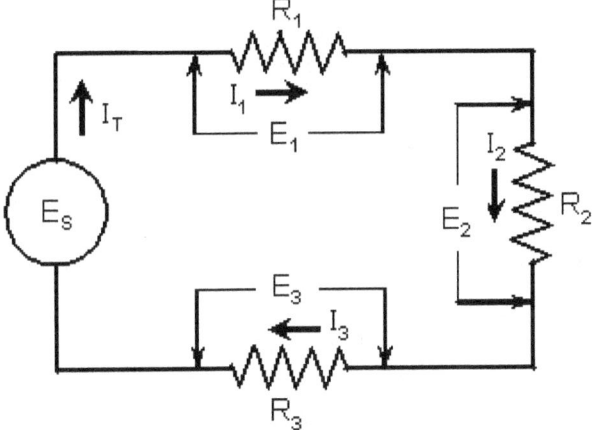

Here are the basic rules of a series circuit.

- Current: In a series circuit the current (I) in amperes is the same everywhere in the circuit.

  IT= I1 = I2 = I3

- Voltage: The total voltage of the circuit is the sum of the voltages across each of the resistors in the circuit.

  VT = V1 + V2+ V3

- Resistance: The total resistance of the circuit is the sum of the individual resistors in the circuit.

  RT = R1 + R2 + R3

Parallel Circuit:

A parallel circuit is one in which the elements in parallel have the same voltage but may have different currents. As seen in the diagram below, it has numerous pathways for the current to take:

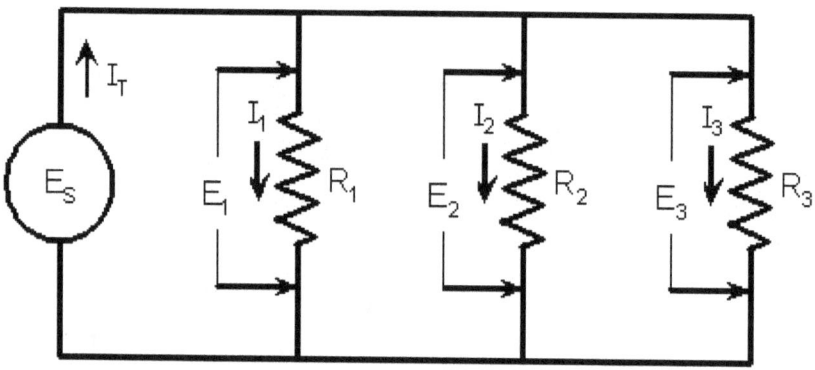

Basic Rules

- **Voltage:** The voltage is same across all resistors in the circuit and it is equal to the supply voltage.

  $VT = V1 = V2 = V3$

- **Current:** The total current (I) in amperes flowing in the circuit is the sum of the currents through each parallel branch of the circuit.

  $IT = I1 + I2 + I3$

- **Resistance:** The total resistance of a circuit where all resistors are in parallel is a little difficult to determine. The reciprocal of the total resistance is the sum of the reciprocals of each resistance. It is very important to note that the total resistance is smaller than the smallest resistance in the circuit. If all the resistors are of the same value, then just divide the resistance by the total number of

  $$\frac{1}{RT} = \frac{1}{R1} + \frac{1}{R2} + \frac{1}{R3}$$

  resistors.

A good method for calculating total resistance is to apply a voltage to the circuit and then calculate total current flow. The total resistance is then calculated by dividing the voltage by the total current flow using

Ohm's equation.

Open Circuits:

An open circuit is one in which the path has been interrupted or "opened" at some point, preventing current from flowing. An open circuit is sometimes known as a short circuit.

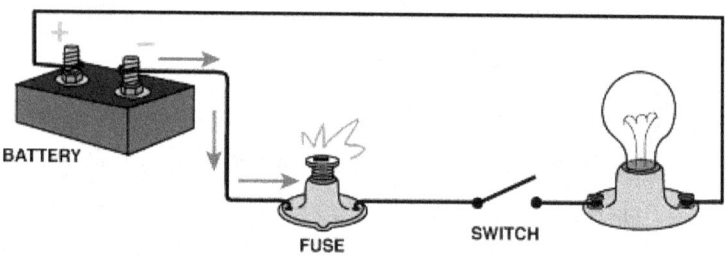

An open circuit might be deliberate or unintentional. An purposefully open circuit would be the circuit to the room's turned-off lights. Because the switch is in the "off" position, which "opens" the path via which power would ordinarily flow, there is no closed path available for energy to flow to the lights.

An example of an unintentionally open circuit is when a circuit breaker trip owing to too much current on the circuit and shuts it off. The circuit breaker or fuse "opened" or "tripped" the circuit, as the electrical industry word goes.

This was accomplished by "opening" the circuit breaker switch.

Short Circuits:

A short circuit is one in which the electricity has found an alternate path back to the source without passing through an appropriate load.

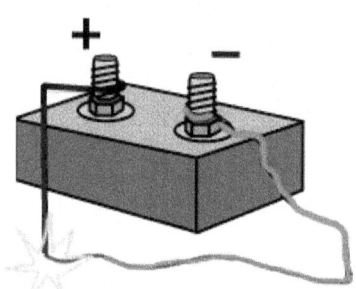

You may readily illustrate this by attaching a thin length of wire to both the positive and negative terminals of a small battery. The wire will rapidly heat up and perhaps melt. This high amperage reflects a dangerous situation in most circuits, as it could ignite a fire or electrocute someone.

Power Circuits:

A power circuit is any circuit that delivers power to electrical loads. Power circuits frequently transport high voltages and are made up of incoming main power, a motor starter, and the motor itself. This may appear to be a simplistic concept, but it is critical to distinguish between power and control circuits since they serve different functions.

Control Circuits:

A control circuit is a type of circuit that uses control devices to govern current flow to detect when loads are activated or de-powered. Power circuits often carry higher voltages than control circuits. Consider a huge industrial engine with 600 horsepower running a water pump. The motor is powered by a 680V high-voltage power supply. When this motor is turned on, it must draw enough current to move the water, and it is customary for a motor to draw approximately six times its regular working current for a short amount of time. This can be inconvenient.

- The first issue is the operator's ability to close the switch safely.
- The second issue is that when the operator opens the switch to turn off the motor, the energy continues to attempt to complete the course. As the switch is opened, this will cause an arc between the contacts. Not only is this arcing harmful, but it also destroys the switch by severely scorching the contact points.

A control circuit ensures that the motor starts and stops safely for both the operator and the equipment. The thermostat to the air conditioner is a popular control circuit example. The thermostat is part of a low-voltage control circuit that regulates a relay that turns on and off the power circuit to the air conditioning   compressor.

POWER (P)

<u>For DC circuits</u>, power (watts) is simply a product of voltage and current.

P = Volts x Amps

<u>In AC circuits</u>, the formula holds true for purely resistive circuits; however, for the following types of AC circuits, power is not just a product of voltage and current.

- <u>Single-phase power:</u>

  Power (kW) = Volts × Amps × power factor

- <u>3-phase power:</u>

  Power (kW) = 1.73 × Volts × Amps × power factor

Power consumption:

The total amount of energy consumed is calculated by multiplying the power by the time the load is turned on. This is most commonly expressed in "Kilowatt Hours" (or kWh) and is what the power company generally uses to calculate your bill. A kilowatt-hour (kWh) is 1,000 watts used for one hour. As an example, a 100-watt light bulb operating for ten hours would use one kilowatt-hour.

Power factor:

Power factor is defined as the ratio of real power to apparent power. The maximum value it can carry is either 1 or 100(%), which would be obtained in a purely resistive circuit.

Power factor = True power / Apparent power

- **Apparent power (VA)** is the product of voltage and ampere, i.e., VA or kVA is known as apparent power. Apparent power is total power supplied to a circuit inclusive of the true and reactive power.
- **Real power or true power** is the power that can be converted into work and is measured in watts.
- **Reactive power:** If the circuit is of an inductive or capacitive type, then the reactive component consumes power and cannot be converted into work. This is known as reactive power and is denoted by the unit VAR.

Relationship between powers:

- Apparent power (VA) = V × A
- True power (Watts) = VA × $\cos\varphi$
- Reactive power (VAR) = VA × $\sin\varphi$

A circuit's power factor is a value that can vary from zero to one and is only found in alternating current circuits. In a circuit, inductance or capacitance can cause the voltage sine wave and current sine wave to diverge and not reach a peak or zero at the same moment. When this occurs, the power factor falls below one. The greater the misalignment of current and voltage, the lower the power factor of the circuit. Since electric motors have a high inductance, their power factor is typically less than 1.0. The voltage and current of an incandescent light bulb or a resistance type electric heater are in phase, and the power factor is 1.0. As a result, determining the current drawn by a light bulb is as simple as dividing the wattage by the voltage.

In the case of a three-phase circuit, the load is supplied by three conductors rather than two wires as in the case of a single-phase load. The current in one cable supplying the three-phase load is 120E out of phase with the current in the other wires. The square root of three, 1.73, is a factor that takes all of this into account. By comparing the previous two calculations, you can see that if the power, voltage, and power factor are all the same, a three-phase load draws less current than a single-phase load of the same wattage.

POWER SUPPLY

Single Phase Power:

Single phase power circuits have two wires, one for power and one for neutral. The majority of residential houses are powered by single-phase electricity. The power company connects three wires into a home, two hot wires and one neutral wire. The neutral is actually a center-tapped transformer feed. Voltage measured across both hot wires

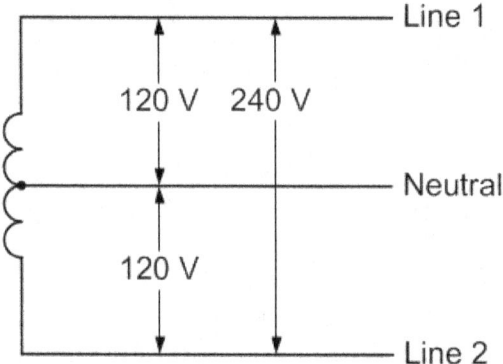

is 240 VAC and the voltage measured from any hot to neutral is 120 VAC (split-phase). Some people mistakenly believe that a 240 VAC circuit is "two-phase", but it's actually the full phase of a single-phase circuit whereas the 120 VAC feeds are half-phase (split-phase).

The selling feature of 240-volt power is that it's twice as powerful and twice as efficient as 120 volts (allows you to run higher–wattage appliances at half the amperes).

- Line 1 to neutral and Line 2 to neutral are used to power 120 volt lighting and plug loads.
- Line 1 to Line 2 is used to power 240 volt single phase loads such as a water heater, electric range, or air conditioner.

3 - Phase Power:

Three Phase power has three power conductors (120V, 120V, 120V) out of phase with one another and one neutral conductor. For our purposes let's consider a 3 Phase 4 Wire 208Y/120V power circuit. This arrangement provides (3) 120V single phase power circuits and/or (1) 208V three phase power circuit.

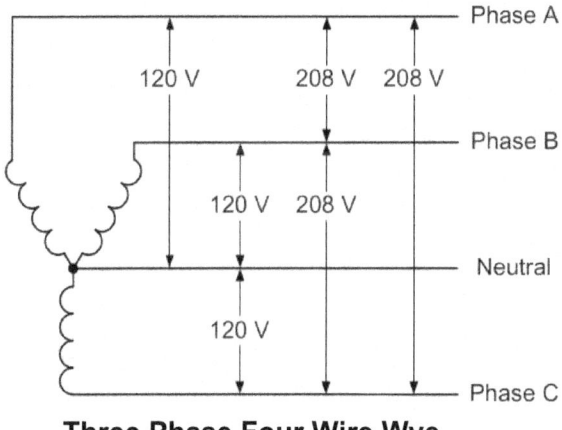

Three Phase Four Wire Wye

The most common commercial building electric service in North America is 120/208 volt wye, which is used to power 120 volt plug loads, lighting, and smaller HVAC systems.In larger facilities the voltage is 277/480 volt and used to power larger HVAC loads.

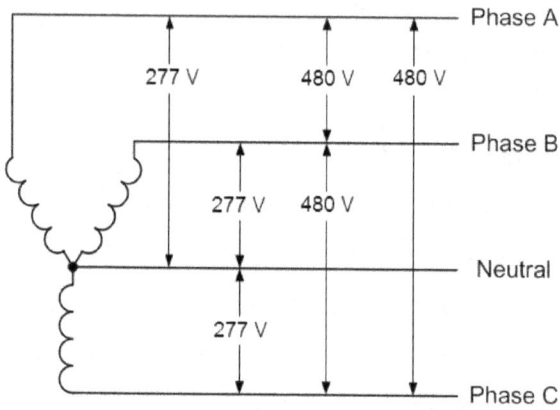

Three Phase Four Wire Wye

Three Phase Three Wire Delta:

Used primarily in industrial facilities, 3-wire delta configuration is used to provide power for three-phase motor loads, and in utility power distribution applications. Nominal service voltages of 240, 400, 480, 600, and higher are typical.

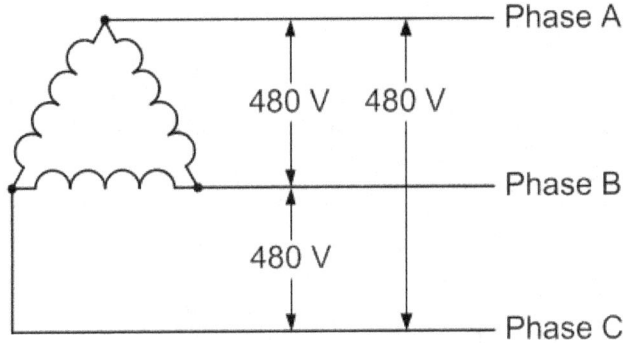

Three Phase Three Wire Delta

CONDUCTORS

The term "conductor" refers to anything that allows or conducts the flow of electricity. Electricity follows the route of least resistance, and certain materials permit more free flow of energy than others. Although aluminum and copper-clad aluminum wires are also used, copper is a good conductor. Insulators are materials that prevent electric current from flowing. Wood and plastic are excellent insulators.

Electric wires:

Electrical wires are conductors that are sized in two different systems: the American Wire Gauge System (AWG) and the Thousand Circular Mill system (KCMIL), which was known until recently as (MCM). Both systems designate wire size based on their diameters or cross sectional areas. The American Wire Gauge system is used to refer to relatively small wires.

AWG sizes: 16gauge to 0000 (4/0). Size increases as the number decreases.

- 16 gauge smallest = .05"
- 4/0 largest = .46"
- 14 gauge copper wire is the minimum gauge allowed in construction.
- MCM: cable larger than 4/0. Sizes are 250, 300, 400, 500 (measurement representing square of cable diameter in thousandths of an inch).

Ampacity:

The current carrying capacity of a particular wire is dictated by its "ampacity" - how many amps it can handle. Ampacity is a function of the cross-section area or diameter of the wire and its material type. Larger diameter wires have larger cross section areas and can safely carry more electrical current without overheating. The maximum ampacity for different types of wires is reported in the electrical codes and standards and tabulated based on the size of the wire, temperature application and the insulation type for the particular wire.

Materials:

Basic conductors: copper and aluminum

- Aluminum conductors must be larger to carry same amperage but are lighter and have lower installation cost. Furthermore, they require special installation because joints loosen and oxides form, causing resistance and overheating.
- Copper conductors are cost effective in small and medium sized wires.

Conductor Resistance:

The resistance of a conductor is determined by the material used. Because copper is a better conductor than aluminum, the resistance of an aluminum wire is greater than that of a copper wire of the same size and length. Wire resistance increases with increasing temperature and lowers with decreasing temperature. A wire's resistance will change by around 8% for every 25°C change in conductor temperature, as a rough guess.

The resistance of a wire is proportional to its length. As the cross-sectional area of a wire grows, so does its resistance. A size 10 AWG wire, for example, has almost four times the cross-sectional area of a size 16 AWG wire.

CHAPTER-2: POWER DISTRIBUTION ELEMENTS

When electrical energy is generated, it is instantly changed and transmitted to the consumer via a network of wires, substations, and transformers. Electricity is generated at a relatively low voltage of 6.6 kV, 11 kV, or 33 kV (r.m.s.) depending on generator design and is stepped up to values as high as 69 kV to 132 kV or more before being fed into transmission lines using step up transformers. The cables themselves are often composed of aluminum (low resistivity) on a steel core (strength) and supported by ceramic insulator strings on steel pylons. The Grid System is made up of this network of transmission lines and pylons.

This voltage is stepped down to the needed voltage at a Substation to meet the power needs of a specific region. This voltage is then decreased further by smaller Substation Transformers. A Tertiary Grid System (voltage is 11 kV at this point) has multiple branches that feed small local Sub-Station Transformers, where the voltage is finally dropped to 480/240V levels.

The energy from generated electricity undergoes several voltage and direction adjustments from the generating plant to the final destination. Each modification necessitates skilled design and implementation in order to give the user with the least expensive, most reliable energy available.

Substation, Switchyard, Switchgear:

- Substation is place where high voltage is stepped down via step-down transformers. Most substations include one or more transformers and switchgear to control the flow of electricity into and out of the substation. When we define substation, we call it 220/33kV, 100MVA substation. Substation can be indoor as well as outdoor.

- Switchyards deliver the generated power from power plant at desired voltage level to the nearest grid. When we define switchyard, we call it 220kV switchyard, 33kV switchyard. Switchyards are generally in open yards. A large fence is generally used to keep the public out of switchyards and open substations since high voltage electricity is dangerous.

- Switchgear is a combination of switching devices such as electrical disconnect fuses and/or circuit breakers, relays and other electrical controls used for control, metering (measurement) and switching. Switchgear is used both to de-energize equipment to allow work to be done and to clear faults downstream.

Power Distribution:

It is advantageous to transmit power at high voltage for economic and efficiency reasons. The power generated: $P = V \times I$ (Watts). However, certain losses occur during the conducting process which amount to: $I^2 \times R$ (Watts).

Power Available $= (V \times I) - (I^2 \times R)$ (Watts)

As a result, these losses must be reduced to a minimum. This indicates that either I or R must be kept as small as possible. To maintain R, tiny metals with low resistivity (copper or aluminum) could be used in big diameter cables, but a thick cable over long distances would be prohibitively expensive. As a result, removing "current, I" is preferable. This is possible, but in order to keep the same power output, the voltage must be increased. A Transformer is utilized to do this.

Transformers:

The transformer is the heart of the substation. The transformer changes the relationship between the incoming voltage and current and the outgoing voltage and current. It consists of two separate coils of wire wound around a laminated steel core. When an alternating current is passed through one coil of wire the current flow creates a magnetic field around the coil. The second coil of wire, usually wound directly over the first coil, is within the magnetic field created by the current in the first coil. Since the current in the first coil is alternating back and forth, the magnetic field is in constant motion. The moving magnetic field induces a current flow in the second coil of wire. The relationship between the voltage of the first coil and the voltage of the second coil is directly proportional to the number of turns of wire on the first coil as compared to the number of turns of wire on the second coil. If the first coil (called the primary winding) has twice as many turns as the second coil (called the secondary winding), then the voltage of the secondary winding is only half that of the primary winding. This is also called the "Turns Ratio". Cold rolled grain oriented (CRGO) steel is used as the core material to provide a low reluctance, low loss flux path. The steel is in the form of varnished laminations to reduce eddy current flow and losses on the account of this.

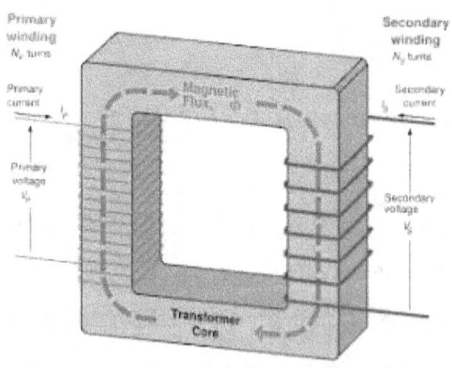

There is a very simple and straight relationship between the potential across the primary coil and the potential induced in the secondary coil. The ratio of the primary potential to the secondary potential is the ratio of the number of turns in each and is represented as follows:

$N_1/N_2 = V_1/V_2$

Where:

- N_1 = Number of turns in primary
- N_2 = Number of turns in secondary
- V_1 = Voltage in primary
- V_2 = Voltage in secondary

Another important fundamental principle of transformers is that when the transformer is loaded, the current is inversely proportional to the voltages and is represented as follows:

$N_1/N_2 = V_1/V_2 = I_2/I_1$

- N_1 = Number of turns in primary
- N_2 = Number of turns in secondary
- V_1 = Voltage in primary
- V_2 = Voltage in secondary
- I_1 = Current in primary

- I2 = Current in secondary

There are certain losses due to heating in an actual transformer, and this does not hold perfectly turn. This relationship is considered to be valid for the purposes of installing transformers and wiring, as well as overcurrent protection for transformers, because it depicts the worst-case scenario. It is vital to note that if the secondary voltage is only half that of the primary voltage, the secondary current must be double that of the primary current to make both sides of the equation equal.

SECONDARY POWER DISTRIBUTION

Electricity leaves the distribution substation and is distributed to switchboards via three phase distribution wires. The supply or intake cable may enter the structure via an underground duct or an overhead supply. Because all of the electrical service is hidden, a subterranean supply is chosen. The supply connection is terminated in the board's fused sealing chamber, which holds an electric meter that records the amount of electricity consumed in kilowatt-hour units as well as peak demand. Switchboards divide a huge block of electricity into distinct circuits, each of which is managed and protected by the switchboard's fuses or switchgear.

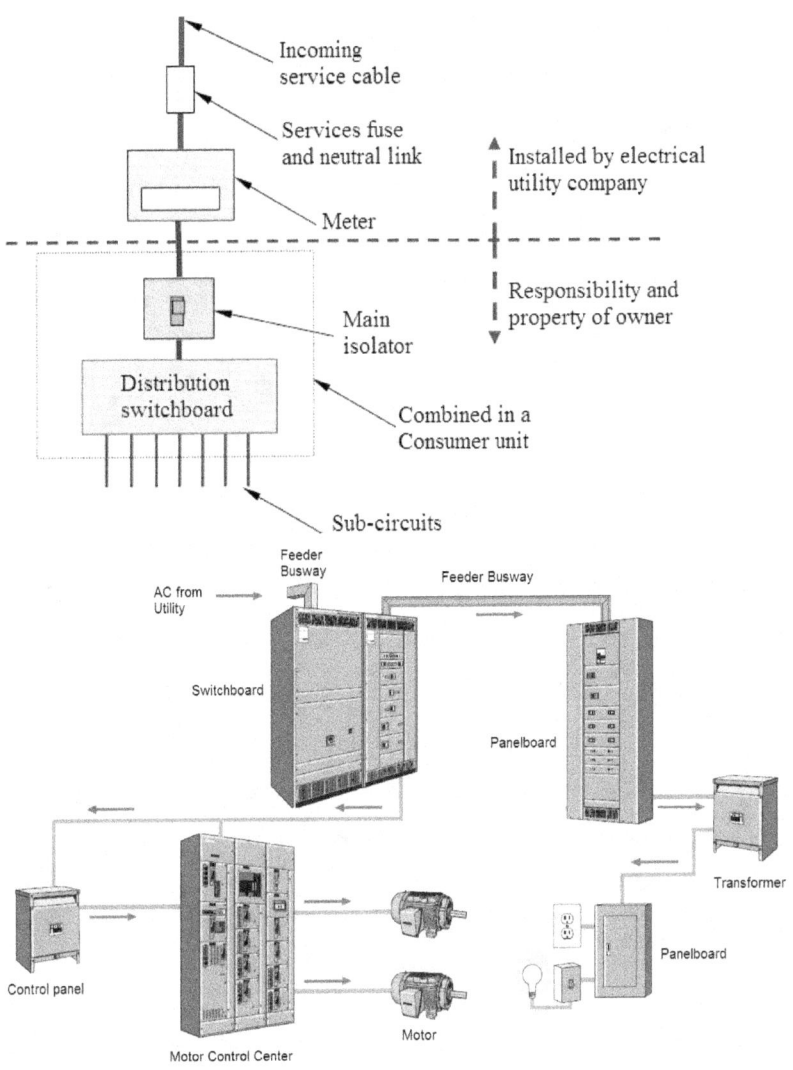

Electric Switchboard:

A switchboard is a device that transfers electricity from one source to another. A switchboard's role is to divide the primary current supplied to it into smaller currents for further distribution, as well as to provide switching, current protection, and metering for these multiple currents. Switchboards, in general, transfer power to transformers, panelboards, control equipment, and, eventually, system loads, MCC.

Inside the switchboard there is a bank of busbars (flat strips of copper or aluminum) to which the switchgear is connected. These carry large currents through the switchboard, and are supported by insulators. A switchboard may include a metering or control compartment separated from the power distribution conductors.

Panel Board:

A panel board is an enclosed assemblage of circuit breakers, fused switches, or, in rare cases, fuses connected to a bus. Branch circuits provide power to receptacles, switches, fixtures, and appliances located throughout the structure.

- A 120 volt circuit consists of one hot conductor

and one neutral conductor. The hot conductor originates at the breaker or fuse connected to one of the hot bus bars.

- A 240 volt circuit requires both hot bus conductors, so it originates at a breaker or fuse connected to both hot bus bars.

While both panel boards and switchboards handle power control and circuit protection, there are significant variations between these systems. A panel board must be installed in or against a wall, whereas some switchboards must be installed away from a wall to provide installation and maintenance access to the rear of the device. The quantity of power regulated by each sort of system, however, may be the most significant variation. In general, switchboards can be adjusted to contain larger circuit breakers or switches to handle higher currents. This also implies that switchboards can be more complicated and include a wider range of devices.

Motor Control Center (MCC):

A motor control center is a factory-built assembly comprising one or more vertical metal cabinet parts with a power bus and associated switchgear for managing a set of motor feeders. To name a few, each segment may include compartmentalized starters, feeders, transformers, adjustable frequency drives, and panel boards. Most starters and feeders are plug-in, depending on size. These units are normally rated at 600 volts alternating current.

They are modular in design, compartmentalized, fixed or draw-out, and can be used indoors or outdoors. Motor control centers provide wire paths for field control and power cables and can be configured with a variety of options such as separate control transformers, pilot lamps, control switches, extra control terminal blocks, various types of bi-metal and solid-state overload protection relays, and various classes of power fuses or circuit breakers.

A Typical Motor Control Center

A motor control center can be supplied ready to connect all field wiring by the customer, or it can be an engineered assembly with internal control and interlocking wiring to a central control terminal panel board or programmable controller. MCC enclosures are graded according on their ability to withstand the intrusion of solid particles such as dust and vermin as well as liquids such as water, oil, and so on.

NEMA rated motor control centers are typically the industry standard in the United States, Canada, and much of Mexico. Motor control centers that are IEC rated are typically found in Europe, Asia, Australia, and Brazil. The standards describe enclosure rating, busbar, short circuit, and seismic rating, as well as wire class drawings.

Control Panel:

A control panel is made up of a controller. The controller could be a PLC, a DCS, a relay, or something else. It sends a digital signal input signal to the MCC panel, which causes the motor to start. The control panel is powered by a PLC/DCS program or relay logic. Normally, instruments are linked to the control panel. The control panel also has interlock indications. Single panels are now increasingly popular due to the use of separate control and MCC panels.

Busbars:

Busbars are conductors that connect two or more circuits together. The transformer's secondary is often suspended from the ground on insulators attached to a stiff metallic bar, which is subsequently tapped at many spots to give electricity to distribution feeders. Engineers call an energized bar a "busbar" or simply "bus." Because of its stiffness and higher current carrying capability, a busbar is utilized instead of a wire. Busbars are made of E91 E grade aluminum or electrolytic grade copper. The following care should be followed when selecting and mounting busbars:

- Busbars are sized considering system fault current and for specified load current with respect to ambient temperature and final working temperature.
- Busbars should be insulated and properly supported with adequate clearances between phases-neutral-earth.
- For MCC, the main bus should be rated to carry 125% of the largest motor running in addition to 100% of the full load rating of all the other motors operated at the same time. Allowances should take into account motor duty cycle and demand factor. Allowances should be made toward future loads.

Busbars Building Electrical Distribution Systems:

A single line diagram (SLD) is a diagram that represents all of the components and identifiable aspects of an electric circuit and is widely used to examine a building's electrical system.

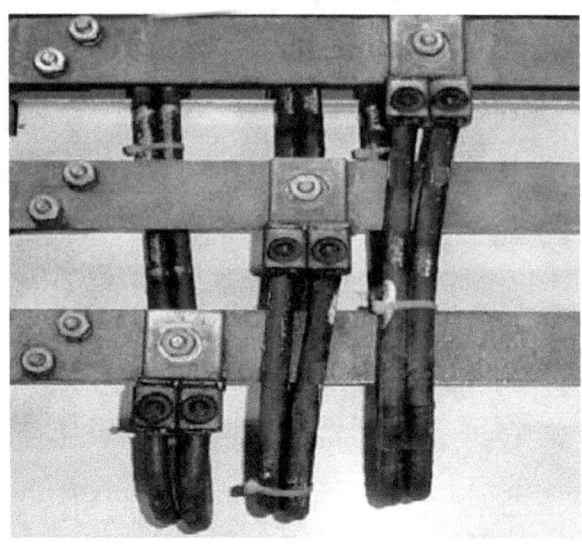

A single line diagram (SLD) is a circuit diagram in which "one-line" represents the three phases of a three-phase power system. A properly designed one-line diagram not only indicates the ratings and sizes of electrical equipment and circuit wires, but it also illustrates an electrically correct distribution of power with respect to current flow from the power source to the downstream loads or panel boards. Circuit breakers, transformers, and other electrical components

Standardized schematic symbols represent capacitors, busbars, and conductors. The graphic elements do not represent the true size or position of the electrical equipment. The single line diagram can be read from the top to the bottom or from left to right. The principle is illustrated in the example below.

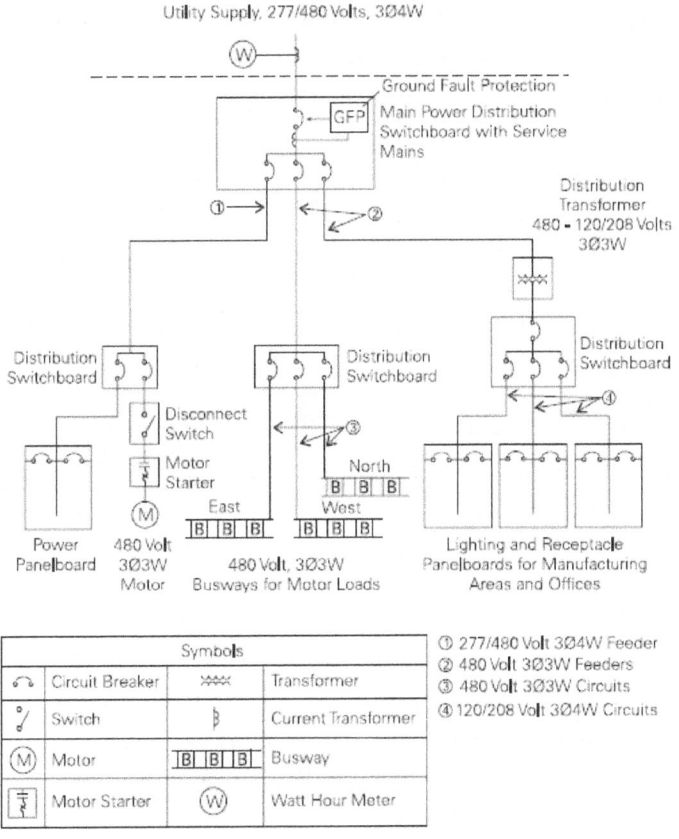

Utility Supply, 277/480 Volts, 3Ø4W

Ground Fault Protection

GFP Main Power Distribution
Switchboard with Service
Mains

Distribution
Transformer
480 - 120/208 Volts
3Ø3W

Distribution
Switchboard

Distribution
Switchboard

Distribution
Switchboard

Disconnect
Switch

Motor
Starter

North
|B| |B| |B|

East
|B| |B| |B|

West
|B| |B| |B|

Power
Panelboard

480 Volt
3Ø3W
Motor

480 Volt, 3Ø3W
Busways for Motor Loads

Lighting and Receptacle
Panelboards for Manufacturing
Areas and Offices

Symbols			
⌒	Circuit Breaker	⋙	Transformer
⌀/	Switch	β	Current Transformer
(M)	Motor	|B| |B| |B|	Busway
⤵	Motor Starter	(W)	Watt Hour Meter

① 277/480 Volt 3Ø4W Feeder
② 480 Volt 3Ø3W Feeders
③ 480 Volt 3Ø3W Circuits
④ 120/208 Volt 3Ø4W Circuits

According to the diagram, power from the utility company is metered and enters the facility via a distribution switchboard. The switchboard serves as the primary disconnecting device for three independent distribution boards.

1. The feeder on the left feeds a distribution switchboard, which in turn feeds a panel board and a 480V, three-phase, three-wire (3Ø3W) motor.

41

2. The middle feeder feeds another switchboard, which divides the power into three, three- phase, three-wire circuits. Each circuit feeds a busway run to 480 volt motors.

The feeder on the right provides 120/208 volt power to lighting and receptacle panel boards through a step-down transformer. Branch circuits from the lighting and receptacle panel boards power the plant's lighting and outlets. Busway can often be utilized instead of cable/conduit feeders at a reduced cost.

Importance of Single Line Diagrams:

The facility can benefit from single line diagrams in a variety of ways. Components in one-line power diagrams are typically placed in decreasing voltage level order. The highest voltage component is displayed in the drawing at the top right. Building maintenance personnel and electricians rely on one-line diagrams to navigate the electrical system. Start at the component and trace the flow of power backwards through the drawing to determine how power is delivered to it. This procedure is very useful for locating the proper circuit breaker to isolate a component for maintenance. Facility managers can use the information contained in single-line diagrams to significantly improve the performance of service activities.

ELECTRICAL PROTECTION & CONTROL

Circuits are controlled by switchgear, which is designed so that the circuit can be safely operated under normal conditions, automatically isolated under fault conditions, or manually isolated for safe repair. These needs are met through good craftsmanship and the use of appropriate materials such as:

1. Circuit breakers
2. Fuses

3. DIsconnect switches or isolators

4. Capacitors

5. Relays

6. Contractors

7. Starters

PROTECTION AGAINST OVER-CURRENT

Any wire or conductor's amperage (current flow) is limited to the maximum authorized by design. Over-current protection is placed to give an automatic way of interrupting or opening a circuit with load currents greater than their rating, as well as in the event of a fault or short circuit. Circuit breakers and fuses, both rated in amperes, are the most prevalent forms of over-current devices.

Circuit breakers:

Circuit breaker is a generic phrase with an implied meaning: something that breaks a circuit. A circuit breaker is a switching device that protects the distribution line or feeder to which it is connected from overloads and malfunctions. If a circuit overloads, the breaker's internal mechanism triggers the

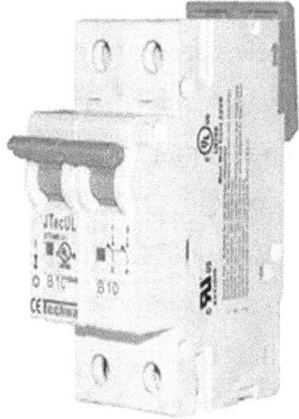

switch and breaks the circuit. Simply turning the switch will reset the circuit breaker. A circuit breaker may withstand short-term overloads (such as the high initial current required in starting a motor, such as a fan motor) without tripping, but it cannot withstand long-term overloads. After the source of the problem has been identified and corrected, power can be quickly restored by switching the circuit breaker. A circuit breaker is extensively employed to protect an entire electrical installation against a big short circuit current or current drawn in excess of its rated capacity since it is designed to make as well as break a large quantity of power.

Fuses:

A fuse is a device that protects switchgear equipment and cables from overcurrents. When a fuse element bursts, the circuit breaks and an arc forms between the breaking points, releasing a huge quantity of heat that can harm neighboring equipment and set wires and cables on fire.

To avoid this, fuse elements are housed in a sturdy and non-flammable housing (usually ceramic) filled with quartz, so that when the fuse element blows, the sand falls down and covers the live contact. This is referred to as a High Rupturing Capacity fuse (HRC fuse). The High Rupturing Capacity fuse has a significant benefit in that it begins to blow long before the entire short circuit current goes through the fuse element. As a result, an HRC fuse is the quickest acting device to protect equipment from a short circuit.

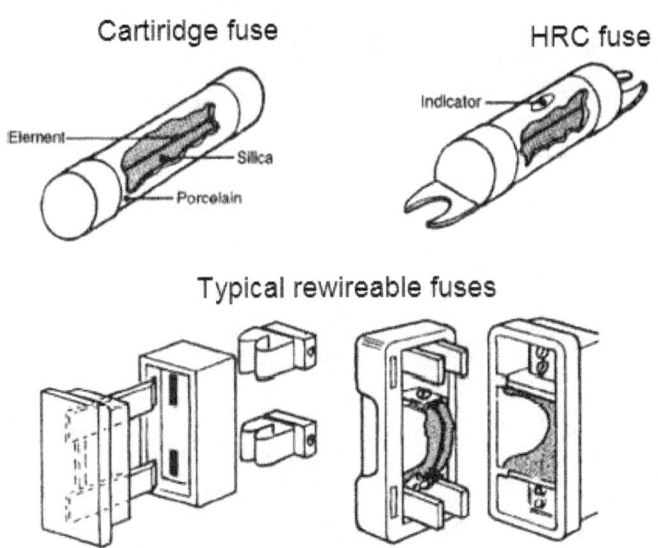

Cartiridge fuse

HRC fuse

Typical rewireable fuses

Isolators and Switches:

An isolator is a mechanical mechanism that must be manually opened in order to disconnect the entire system, one circuit, or one piece of equipment from the live supply. Furthermore, a way of turning off for maintenance or emergency switching must be provided. A switch can offer isolation, but an isolator, unlike a switch, is intended to be opened when the circuit in question is not carrying current. Its objective is to guarantee the safety of anyone working on the circuit by rendering those components that are normally live inactive.

It should be noted that an isolator can only establish the circuit in the absence of a load. It cannot establish or break any load current, whereas a switch can make or break an electrical circuit while operating at rated load current. One device may provide both isolation and switching if its properties match the requirements for both tasks.

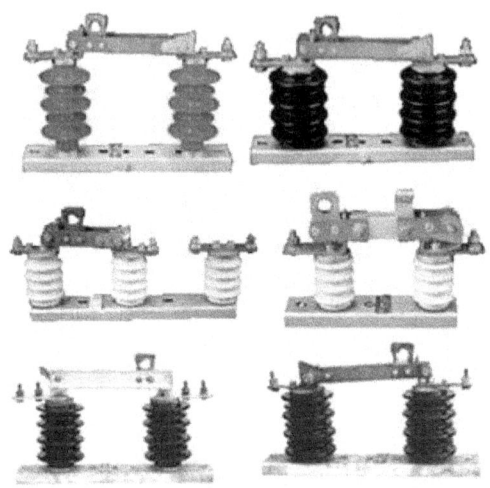

Disconnecting Switch

Oil Cutout (Oil-Filled Cutout):

A cutout in which all or part of the fuse support and its fuse link or disconnecting blade are completely immersed in oil, with complete immersion of the contacts and the fusible portion of the conducting element (fuse link), so that arc interruption occurs under oil by severing the fuse link or opening the contacts.

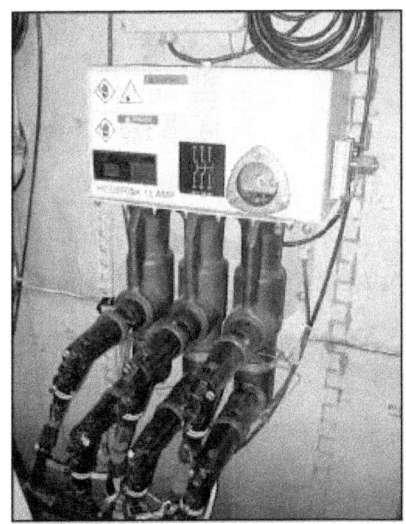

Oil Cutout

Ground fault interrupter (GFI):

A device that detects small current leaks and disconnects the circuit's hot wire. It may also be a component of a circuit breaker.

Capacitors:

Capacitors help to modify the power factor and voltage, allowing for more efficient electricity distribution. They can be controlled remotely and swapped in and out of the system as needed. Some capacitor banks are temperature controlled, so that if the temperature rises beyond a specified threshold, the capacitor bank is automatically switched into the circuit. Others are programmed to automatically connect and disengage from the system at pre-determined times, generally matching to the operation of a large factory or another load with a low power factor.

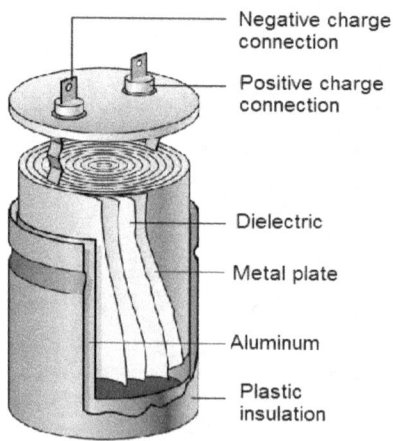

Negative charge connection

Positive charge connection

Dielectric

Metal plate

Aluminum

Plastic insulation

Relays:

Relays are small, extremely fast-acting automatic switches that are used to protect an electrical system from faults and overloads. It is commonly an electromagnetic device with a coil. When this coil is powered, a magnetic field is formed that operates a mechanical switch. When a relay detects a problem, it instantly sends a signal to one or more circuit breakers to open, or trip, safeguarding both the relay and human life from harm. Relays can also be activated via communication lines to open or close circuit breakers based on a command.

Electromagentic Relay

In HVAC systems, the following protective relays are typically used:

Overload/overcurrent relay (thermal and magnetic), phase reversal relay, short circuit protection relay, earth fault relay, thermistor relay, and special motor protection relay (for protection of higher HP rated motors).

Contactors:

A contactor is an electrically operated switch that can be made to switch on or switch off a motor, heater bank, capacitor bank,etc. directly or by a remote controller such as a thermostat, humidistat, timer, pilot devices or any other protective devices. It consists of 2, 3 or 4 power contacts and some auxiliary contacts.

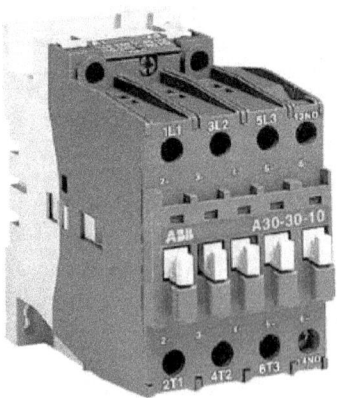

Contactor

Although a switch, a contactor is designed to interrupt (making & breaking) electric current repeatedly and frequently, due to the simplicity of its mechanism and contact design. When a contactor breaks the current, an arc is established across the contacts where the circuit is broken and a good amount of heat energy is generated. This increases when the frequency of breaking the current increases resulting into welding of contacts, or fusing, and contactor failure.

Motor Starters:

A starter is a device which connects with the motor in series to limit the in-rush current at start and establish a starting torque to get the motor to its rated load speed. A starter consists of:

1. A magnetic contactor;
2. A disconnect device: fuse or circuit breaker to protect motor from drawing excess current under sustained overload conditions;
3. An overload relay which may be bi-metallic, melting alloy or solid state; and
4. A control circuit consisting of pilot devices, relays, timers, and PLC's.

It may also include step resistors, disconnects, reactors, auto-transformers or other hardware to make it a more sophisticated starter for large motors.

Motor Starter

A motor starter is designed to provide one or more of the following functions:

- Start, accelerate and stop a motor repeatedly, quickly, safely and dependably;
- Protect motor against operational overloads; and
- Disconnect supply to motor in case of under-voltage/no voltage, if there is danger to operator or to the machinery due to automatic restarting of motor on restoration of full/balanced voltage.

The starter can also perform the following functions with suitable additional devices:

- Protect motor against severe unbalance in voltage and current;
- Limit in-rush/starting current wherever called for;
- Protect starter components/installation from short circuit fault; and
- Provide remote operation facility.

Types of Starters:

Starters are divided into two types:

1. Full voltage or across-the-line or direct on line (DOL) starters where the motor load is directly applied to the line voltage.
2. Reduced voltage or assisted type starter where the motors load is initially applied to a reduced voltage and later to the line voltage. The following reduced voltage starters are commonly used in ACR Systems:
   - Star delta starter
   - Auto transformer starter
   - Part winding starter
   - S ft starter and variable frequency drive

Commonly used starters in HVAC systems:

1. **Direct Online Starter:** DOL starters have the lowest initial cost and since full voltage is applied the motor produces maximum starting torque and the load accelerates fast. Since the motor draws a high starting current to the extent of 600% to 800% of its full load current, larger HP motors will cause a severe dip in the supply system voltage.

2. **Star Delta Starter:** In this starter, one end of the motor windings is initially shorted together and supply is fed to the other end of the motor windings. Thus, even though full voltage is applied to the motor terminals, the effective voltage applied to the windings becomes only 1/3, i.e., 57.7% of the rated voltage. After a preset time, the shorting done at one end of the winding is removed and motor receives full voltage. The reduced supply voltage restricts motor starting current to $(0.577)^2$ i.e., 1/3rd or 33% of full load starting current. Starting is performed in two steps, hence jerk is minimized.

As the voltage applied in each phase is reduced by 1/3 the torque is reduced by $(1/3)^2$ and hence this starter cannot be used where load requires a high starting torque. The time taken for the motor to accelerate is longer due to reduced voltage input. If timing of changeover from star to delta is not set properly the motor is on full line voltage prematurely and will draw a heavy current thereby defeating the main purpose of restricting the starting current. Closed transition is possible only by adding additional resistor or reactance, timer and power contactor, hence this type becomes further expensive. (In a closed transition the current is not broken when the changeover from star to delta takes place, whereas, in an open transition, not only the current is broken when the changeover takes place from star to delta, but also the motor gets a heavy kick/jerk when full voltage is applied).

3. **Auto Transformer:** It is basically a reduced voltage starter in which an auto transformer is used to supply reduced voltage to the motor windings initially. At start, supply from mains is fed to the motor windings through one of the taps on the auto transformer. This reduced supply voltage caused the motor to draw much lower current as compared to the current drawn by a DOL starter. Once the

57

motor reaches almost the rated speed, the auto transformer is disconnected from the circuit and full supply voltage is fed directly to the motor. Thus the motor draws full load current and produces full load torque.

An auto transformer is generally provided with 3 tappings of 50% or 57.7%, 65% and 80% voltages. The corresponding starting current and torque at these voltages are:

Taps at	Starting Current	Starting Torque
50%	25%	25%
57.7%	33%	33%
65%	42%	42%
80%	64%	64%

This type has an advantage of reduction in starting current and torque. Being a two-step starter, starting jerk is minimized compared to DOL. With a minimum of 3 voltage tapping's available in an auto transformer, by selecting appropriate voltage taps, the starter becomes flexible to obtain different starting torque at the site to match different load conditions. Closed transition is possible. Having more costly components like auto transformer, contactors, timer etc., makes it more expensive compared to DOL or star delta starter. It is larger in dimensions than DOL or star delta starter. If the timer for changeover from auto transformer to full supply voltage is not set properly, the motor will changeover to full supply much earlier and it will function as a DOL starter and draw much higher starting current, thereby defeating the basic purpose of current reduction.

4. **Part Winding Starter:** This is one of the simplest types of reduced voltage starter used to control the in-rush current of motor. The poly phase motor has basically two parallel windings, both wound on the same poles. One set of each winding leads are brought out to the terminals and the other set of winding's leads are shorted either internally or externally.

As the windings are in two parts, the power supply is initially fed to one of the windings, thus the motor draws a low starting current. The power to the second winding is fed after a set time and the motor then draws full load current.

Motors in either 67%/33% or 50%/50% windings can be selected depending upon the starting torque requirement. It is less expensive than ATS or star delta starters. Closed circuit transition is possible. Starting torque is better than star delta. Small sized fuses are adequate compared to Dol.

In-rush current is restricted to a maximum of 65% of starting current depending upon the time the second winding is energized. It can withstand better voltage fluctuations. Limitation in motor selection and starting torque is restricted to a maximum of 45% of the rated torque of motor.

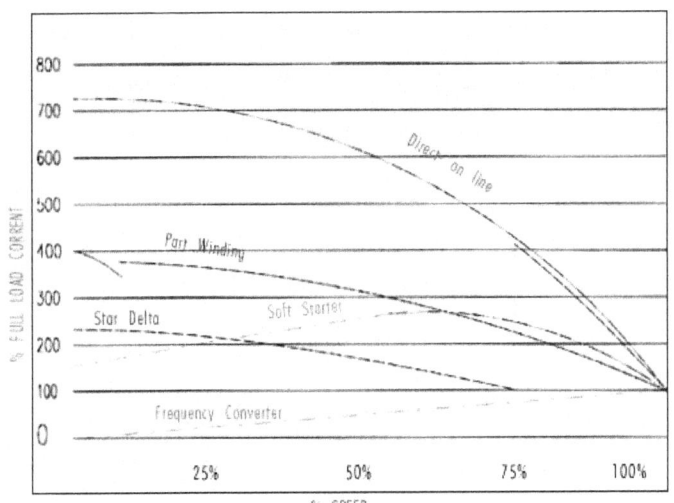

Starting current of various types of starters

5. **Soft Start:** A Soft Start describes a type of motor control that includes a simple solid state power controller. Instead of simply opening and closing the power circuit like a 3 Phase motor contactor, it ramps the motor voltage up or down to turn the motor on and off. A Soft Start is more expensive than other types of motor starters, but provides the added benefit of reducing electrical and mechanical shocks associated with starting and stopping a motor.

6. **Variable Frequency Drive:** A Variable Frequency Drive (VFD) describes a type of motor control that includes an advanced solid state power controller. Instead of simply opening and closing the power

circuit like a 3 Phase motor contactor, or ramping the motor voltage up or down like a soft start to turn the motor on and off, a Variable Frequency Drive (VFD) controls motor speed. A Variable Frequency Drive (VFD) is more expensive than Soft Start, but provides the added benefit of controlling motor speed.

Control Devices:

A control device is a mechanical, electrical, electronic, or thermal switch that turns on or off a circuit breaker, motor starter, heater, and capacitor bank. These include signaling, electric interlocking, and so on. Control devices are classified into two types: manually operated and automatically controlled.

Push buttons and selector switches are examples of manually controlled control devices. Pressure switches, float switches, flow switches, thermostats, thermistor relays, limit switches, protection relays, timers, and auxiliary contactors are examples of automatic control devices.

All manually operated devices are activated by hand, whereas automatic control devices are triggered in response to predefined conditions or requirements.

Remember that control devices can only turn on low-current loads such as an electromagnetic coil in a contactor, small pilot motors in motorized valves and fire dampers, or signaling lamps. They cannot be used to directly interrupt power circuits because they carry enormous currents that can cause the control device to fail immediately.

PROTECTION AGAINST ELECTRIC SHOCK

What causes shock:

Current flows in a continuous closed conduit from the source to a power-using gadget and back to the source. However, electricity does not need to flow via wires to return to the source. It can return through any conducting body, including a human who comes into direct contact with the earth or touches a conductive object, which then enters the earth. You'll also get a shock if you accidently become a link in an electrically live circuit. For example, if a hot wire became dislodged from a light fixture terminal and came into contact with the metal canopy of the light fixture, which is extremely conductive, the fixture would become charged. If you touch the fixture under these conditions, a current leakage or ground fault may occur, providing a conduit to ground for electric current and resulting in a shock. Electric shock protection is accomplished by insulating and placing live parts out of reach in appropriate enclosures, grounding and bonding metal word, and providing fuses or circuit breakers to ensure that the supply is automatically terminated under fault situations.

Grounding (Earthing):

If the circuit had a grounding system, this shock could be avoided. Grounding occurs when one of the conductive wires functioning as part of the circuit path is purposely given a direct passage to the earth, which functions as an effective conductor due to moisture present within the soil. Grounding accomplishes three crucial goals:

- It limits the voltage upon the circuit that might otherwise occur through exposure to lightning or other voltages higher than that for which the circuit is designed.
- It limits the maximum voltage to ground under normal operating conditions.
- It provides automatic opening procedure of the circuit if an accidental or fault ground occurs on one of its ungrounded conductors.

Grounding is normally achieved by connecting one of the circuit wires (often neutral) to the soil or ground by attaching a wire to a ground rod, which is a long copper rod driven directly into the soil. Grounding ensures that all metal portions of a circuit with which one may come into touch are directly linked to the earth, keeping them at zero voltage. A grounding system performs nothing during normal operation. However, in the event of a malfunction, grounding protects against electric shock or fire.

Electrical rules require that all circuits 120 volts or higher have a grounding system. "Grounding" is an American phrase that is equivalent to "Earthing" which is an IEC term.

CHAPTER-3: MOTORS & VARIABLE SPEED DRIVES

In HVAC systems, electric motors are utilized to power fans, pumps, refrigeration equipment, and other processes that require motive (moving) force.

Motors work on a relatively simple basis that most of us have observed when we bring the 'like' poles of two magnets together. A force separates them, as two south poles repel and two north poles repel.

Squirrel cage induction motor:

The Squirrel cage induction motor which is the most common electric motor has four main parts:

1. **Stator:** it is a stationary component made of copper windings that carry current. The stator's coils set up a magnetic field that moves in a circular motion. The stator surrounds the Rotor.
2. **Rotor:** as the name suggests, it rotates. It is caused to rotate under the influence of the

magnetic field of the stator. The rotor tries to keep up with the stator's magnetic field.

3. **Fan:** it is used to cool the motor.

4. **Bearings and seals:** allow a motor shaft to move smoothly and reduce energy losses that would occur through friction. The seals keep dust from entering the motor.

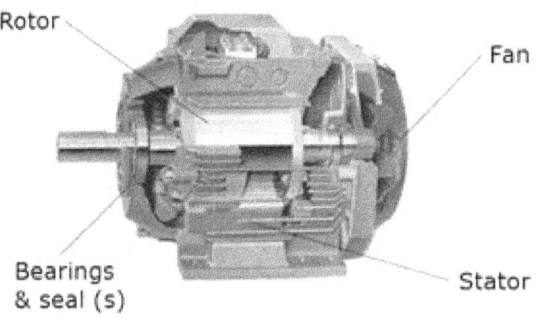

There are several advantages to the induction motor: no brushes or commutator means easier manufacture, no wear, no sparks, no ozone production and none of the energy loss associated with them.

Speed of rotation: synchronous speed:

The synchronous speed of a motor can be determined by the following formula:

$$\text{Synchronous speed} = \frac{120 \times f}{\text{No. of poles}}$$

Where:

- speed is expressed in rpm (revolutions per minute)
- f equals frequency in Hz (hertz)
- poles are an even number, i.e. 2,4,6 etc.

The formula for synchronous speed makes it obvious to see that as the frequency varies, the speed varies in a direct proportion, i.e., if we double the frequency, the speed doubles.

We call the hypothetical speed 'synchronous' speed because it is the maximum speed that would be obtained if the rotor rotated in 'synchrony' with the magnetic field, that is if the rotor kept up with the rotating magnetic field of the stator.

The ideal speed of rotation is determined by two (2) factors:

1. The number of magnetic poles.
2. The frequency of the AC supply.

It is possible to arrange the stator windings in such formations as to provide any number of pairs of poles and so we can offer 2, 4, 6, 8, 10, 12 pole motors. Motors over 12 poles are available if required, but they are not in common use.

The 'slip':

In any AC induction motor the synchronous speed is never achievable, since friction losses in the bearings, air resistance within the motor, and additional drag imposed by the load combine to cause the rotor to lag slightly behind the rotational speed of the magnetic field.

This lagging effect is known as the slip.

The percentage slip varies from one motor to the next. As a general rule of thumb, the larger the motor, the less slip is experienced.

For any given motor the slip will decrease as the load decreases. At no load, the slip may be as little as 0.5%, while at full load and depending on the size of the motor, it can be as high as 5.0%.

Actual speed:

The actual speed is determined by following:

- the slip
- the loading of the motor

It is not surprising to find that the slip of a motor is closely related to the motor's efficiency and, in fact, the full load speed of a motor is a good indicator of the motor's efficiency.

Motor Efficiency:

We all know that to get a motor to do work we need to supply a source of electrical power. In an ideal world, all of the power that is put in would be seen at the output. However, all real systems have losses:

Power losses in induction motors can be grouped into two main components. These are:

Fixed losses, i.e., independent of motor load:

- Iron or magnetic loss in the stator and rotor cores
- Friction and windage loss

Losses proportional to the motor load:

- Resistive (I^2R) or copper loss in the stator and rotor conductors
- Stray loss caused by components of stray flux

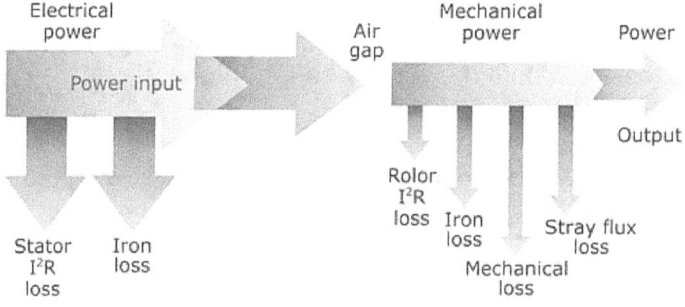

Defining Efficiency:

'Efficiency is the percentage of the power input that reaches the load:

$$\eta = \frac{P_{out}}{P_{in}}$$

Where:

- η is a decimal value; if multiplied by 100, it will give the efficiency as a percentage.
- P_{out} is the output power.
- P_{in} is the input power.

The efficiency rating of an induction motor accounts for the losses in both the stator and the rotor.

- In the *ideal world* an electric motor would be 100% efficient.
- In the *real world* it is more realistic to expect 50% efficiency.
- Low efficiency means higher running costs.
- Not all electric motors are created equal. Some are more efficient than others.

Motor Sizing:

Motors perform best when they are optimally loaded. A considerable drop in efficiency happens at loads of 25% full load or less, and it is at this level that a smaller motor should be seriously considered.

It is essential to note that the amount of power drawn by the motor is determined by the load. The size of the motor does not always correspond to the amount of power drawn. A fan requiring 15 kW, for example, might be operated by a 15 kW motor - in this instance, it is well suited. It may possibly be powered by a 55 kW motor, which would function but would be inefficient. Connecting it to a 10 kW motor, on the other hand, would quickly cause the motor to trip out. This demonstrates the need of understanding the real power drawn by the motor.

High efficiency motor (HEM):

A typical 10kW high efficiency motor might have the following specifications:

- Has an efficiency of 93% whereas standard electric motors have an efficiency of 88%
- Provides a saving of 4.3% in both energy and greenhouse gas emissions

- Recoup the premium paid for a HEM in less than two years

The gap between standard and high efficiency motors becomes increasingly wide as motor size decreases. Many small motors, such as those used in exhaust fans, may have efficiencies as low as 50%.

The figure below shows the variation in efficiency with load for a standard and higher efficiency 7.5 kW motor.

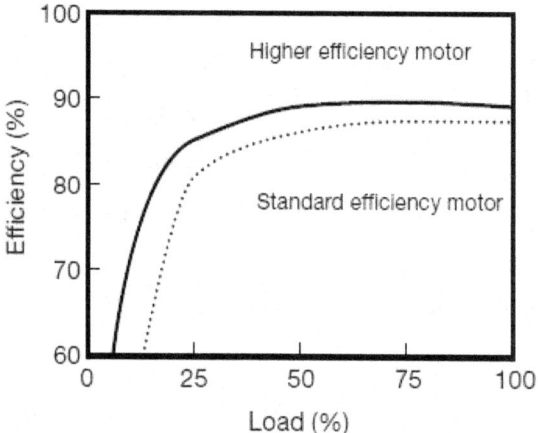

Several factors combine to make a motor energy efficient:

- higher quality low loss laminations for magnetic circuit

- more and better-quality copper in the windings
- better quality insulation
- optimized air gap between the rotor and stator
- reduced fan losses
- closer matching tolerances
- Greater core length

The motor nameplate and motor efficiency tables:

All motors have a metal nameplate fixed to their body. The nameplate gives a number of the motor's characteristics including:
- Brand
- kW or hp (horse power)
- Hz (frequency)
- Amps
- Ambient temperature
- Efficiency (%)
- Voltage (Star / Delta)
- RPM
- cos φ (power factor)

V	Hz	kW	hp	r/min	A	cos φ
690 Y	50	11	15	2935	10.8	0.9
400 Δ	50	11	15	2935	19.2	0.9
660 Y	50	11	15	2930	11.6	0.91
380 Δ	50	11	15	2930	20	0.91
415 Δ	50	11	15	2940	18.7	0.89
460 Δ	60	11	15	3545	16.8	0.9

Example motor nameplate

Rated Motor Power:

The rated motor power is the shaft power, i.e. the useful mechanical power that it can provide to turn the load. However, because the motor itself has losses, the power drawn by the motor at full load is greater than the rated shaft power. For example, at full load, a 30 kW motor that is 92.5% efficient will draw (30/0.925) kW = 32.4 kW.

Motor efficiency calculations:

Example:

A Toshiba 2 pole 11 kW motor has an efficiency of 91% at full load. What is its running cost based on 4000 hours a year at 10 cents per kWh?

Calculation:

First, let's calculate the losses:

11 kW x 9%   (Note: 100% - 91% = 9% = 0.9).

11 kW x 0.09 = 0.99 kW (this is the total of the losses).

Therefore, the total power that must be supplied to the motor will be: 11 kWh + .99 kWh = 11.99 kWh (rounded to 12 kWh). Total power used in the year:

12 kWh x 4000h = 48,000 kWh or 48 MWh. Total cost of running the motor in ONE year:
12 kWh x 4000h x $0.10 = $4,800 a year.

Question:

What would the running costs be for one year if the motor's efficiency was only 81%? 11 kWh x 0.19 = 2.09 kWh (losses: 100% – 81% = 19%)
11 kWh + 2.09 kWh = 13.09 kWh

Total running cost in ONE year:

13.09 x 4000 x 0.10 = $5,236

This means a motor with 81% efficiency will cost an additional $436 a year to run compared with a 91% efficient motor. In ten (10) years this could save $4,360.

Controlling Motors:

Motors must be controlled wherever they are used. The motor controller's functionality and complexity will vary based on the task that the motor is executing.

For example, the most basic method of AC motor control includes turning the motor on and off. This is frequently performed with a motor starter composed of a contactor and an overload relay. The contacts of the contactor are closed to start the motor and opened to stop it. This is done electromechanically by employing start and stop pushbuttons or other pilot devices wired to the contactor. When there is an overload condition, the overload relay disconnects power to the motor. However, an overload relay does not guard against short-circuits. As a result, circuit breakers or fuses are also employed.

More advanced motor controllers can be used to precisely control the speed and torque of the linked motor (or motors), and they can also be utilized as part of closed loop control systems.

A driven machine's position. A numerically controlled lathe, for example, will precisely position the cutting tool according to a preprogrammed profile while compensating for varying load conditions and perturbing pressures to maintain tool position.

One motor starter typically controls one motor. When only a few geographically distributed AC motors are employed, the circuit protection and control components may be housed in a panel adjacent to the motor.

Starting motors:

A starter motor is intended to supply a motor with enough current to produce a starting torque that will allow the motor to reach its rated load speed. One motor starter typically controls one motor. When only a few geographically distributed AC motors are employed, the circuit protection and control components may be housed in a panel adjacent to the motor.

In the sector, there are several strategies for getting started. Each strategy has advantages and disadvantages. The variable speed drive is the most energy efficient.

- **Direct-on-line starting**: stator windings directly connected to supply via contactors
- **Star delta starting**: star and delta connection of stator windings are used
- **Autotransformer starting**: stator windings connected to the supply through an auto

transformer

- **Soft starters**: variable speed drive

A quick comparison of each of the starting methods reveals that the starting current (in-rush current) is considerably higher in some cases than the final current at

Type	In rush current	Torque
DOL	7 x 10 amps = 70 amps	Good
Star/Delta	3x 10 amps = 30 amps	Poor
Auto/Trans	4x 10 amps = 40 amps	Good / Average
VSD	0.5-1.5 x 10 amps = 10 amps	Excellent

a full loaded speed:

Motor Control Circuits:

Understanding the fundamentals of motor control symbols and ladder diagrams is critical for the trip exam. A ladder diagram is a reasonable approach to demonstrate how all of the components of a motor control system work together. The problematic element is that the various control contacts in a device may be scattered over the control diagram. They are assigned a number or a letter. The diagram below depicts various common types of control devices used in a ladder diagram.

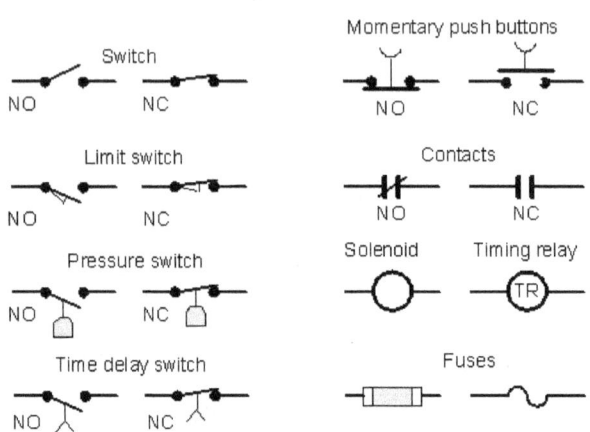

Common control devices found in a motor control ladder diagram

Control devices are classified into two types. One is generally open (NO) and needs to be closed to complete the circuit. The other is generally closed (NC), which means it needs to be opened.

How do you control a magnetic motor starter with a 2-wire control circuit and a 3-wire control circuit:

wire control circuit:

Refer to the diagram below for a simple 2-wire control circuit in which a pressure switch powers a magnetic motor starter. When the control system gives power to a solenoid coil within the motor starter, the primary connections delivering power to the motor close. A circle is commonly used to depict a solenoid coil.

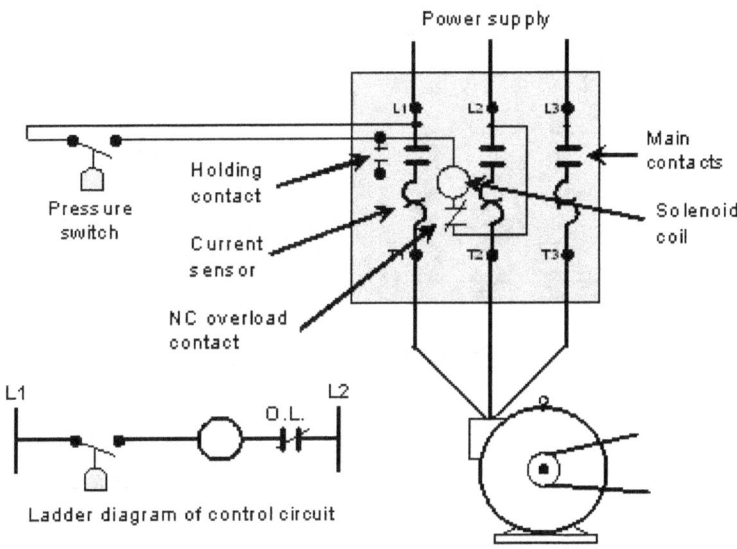

Ladder diagram of control circuit

In the case of a motor overload, one or more normally closed contacts (NC) are inserted in the control circuit to cut power to the solenoid. Typically, some form of gadget measures the motor's current and operates the overload (O.L.) connections. A 2-wire control circuit in a holding contact in the motor starter is   NOT   required   in   this   scenario.

2-wire control circuit:

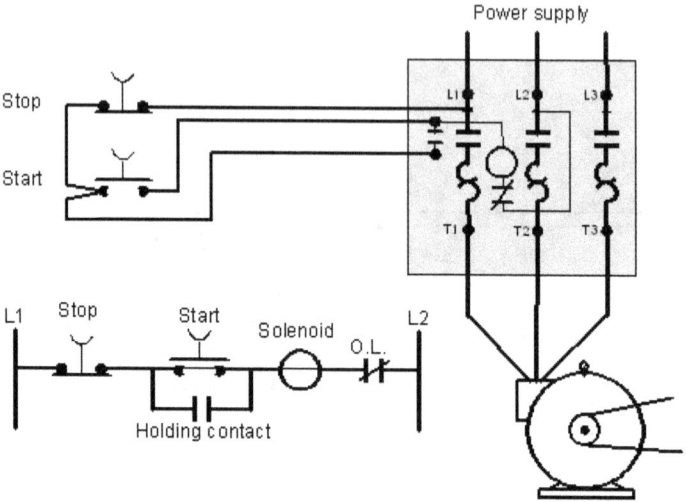

When a motor is controlled by a momentary contact start-stop station, a 3-wire control circuit is employed. It is important to understand at least how to regulate a motor using a variety of momentary control devices. Normally, the stop push button is closed (NC). Pushing this button opens the control circuit and disconnects power from the motor starter's solenoid. The start push button is pressed.

To transfer power to the solenoid coil, the normally open (NO) switch must be pressed. This causes the motor starter's open contacts (NO) to close. When the holding contact in the motor starter is disengaged, it creates a route around the start push button.

84

Any temporary loss of power to the solenoid opens the holding contact, causing the motor to shut off. In the case of this 3-wire control circuit, an operator is required to revive the circuit. When power is restored to the control circuit for the previously described 2-wire control circuit, the motor will start instantly.

Start Stop from 2 different locations:

Refer to the figure below. The stop push buttons are connected in series and the start push buttons are connected in parallel. The holding contact in the motor starter must be connected in parallel with the start push buttons.

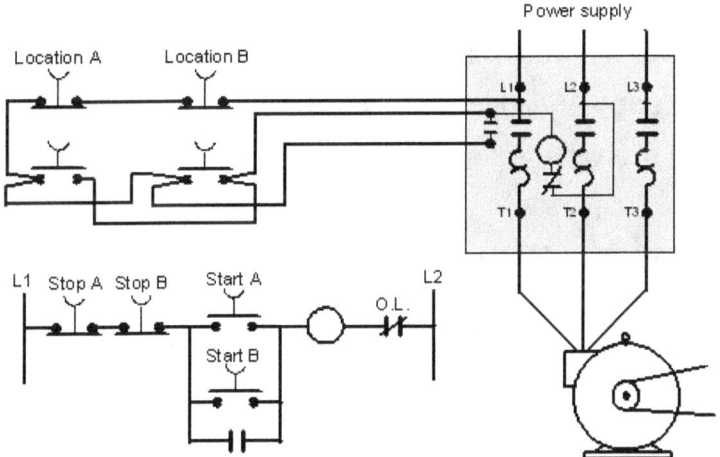

VARIABLE SPEED DRIVES

A typical motor rotates at a nearly constant speed. However, most of the time, the loads being driven do not require the full load power that the motor can provide. Because of the power outage, energy is being squandered, resulting in significant greenhouse gas emissions.

If we could manage the motor's speed so that it more closely fits the requirements of the load, the motor's operating costs would be lower, and greenhouse gas emissions would be reduced.

Variable speed drives (VSDs) or variable frequency drives (VFDs), as they are also known, are widely used in HVAC applications. Variable speed drives (VSDs) enable loads powered by AC induction motors, such as fans and pumps, to run at a variety of speeds.

Advantages of VSD:

Variable speed drives:

- Improve energy efficiency, e.g. savings in excess of 50% can be achieved;

- Improve power factor and process precision;

- Provide other performance benefits such as soft starting;
- Less mechanical stress on the system leading to longer life, lower maintenance overhead;
- Reduce voltage drop on startup, so other systems are less affected, e.g. preventing lights from dimming or other devices from shutting down;
- Over speed capability;
- Eliminate the need for expensive and energy wasting throttling mechanisms, such as control valves and outlet dampers.

Imagine a situation in which you have 10 motors at 10 amps each starting together on a small factory line. The starting currents under various starting methods would be:

- direct-on-line = 1000 amps,
- soft starters = 200 amps, and
- VSDs = 100 amps

What is a variable speed drive:

Recall that the speed of an induction motor is directly proportional to the frequency of the AC supply. If we can vary the frequency of the power supplied to the motor, then we can vary the motor's speed away from its rated synchronous speed. This is exactly what a VSD does!

A VSD is an electronic device that converts constant

frequency AC power input into a variable frequency output. This variable frequency output is used to control motor speed which, as said before, is proportional to the frequency of the VSD's output.

Principles of variable speed drives:

A VSD operates by converting the alternating current main supply to direct current (DC) via a rectifier. After smoothing with a filter, the DC supply is chopped at a high frequency by six transistors to provide variable frequency, variable voltage at the motor terminals, allowing the induction motor to run efficiently at varying speeds. Pulse width modulation (PWM) is a typical method for regulating motor voltage.

Inverter switches with PWM voltage control are used to divide the close-to sinusoidal output waveform into a series of small voltage pulses and regulate the width of the pulses.

Applications:

The most obvious consideration when adding a VFD to a primary mover (such as a fan, pump, or chiller) is that the motor be rated for driving applications (inverter duty motor). Because of the anomalous strains produced to the motor windings and rotor laminations by the VFD, every minor flaw in a motor is accentuated when it is operating at a lower speed/frequency. The outcome can be anything from a very noisy motor to a shorted winding or complete motor failure.

Harmonics are produced by VFD. It is critical that the device include an inbuilt filter and/or isolation transformer that minimizes reflected wave harmonics and prevents them from entering the distribution system.

Electromagnetic compatibility (EMC): You should be aware that motors, cables, and control equipment can release acceptable quantities of electromagnetic radiation. The accidental synthesis, propagation, and receipt of electromagnetic energy is referred to as electromagnetic compatibility (EMC). The purpose of EMC is to achieve proper operation of different devices in the same electromagnetic environment in order to avoid interference effects.

Mitigation of interference, or noise, and thus electromagnetic compatibility is accomplished by addressing both emission and susceptibility issues, i.e., quieting the sources of interference (filtering), making the coupling path between the source and the receiver less efficient, and making potential receiver systems less vulnerable. Manufacturers' equipment and drives must be EMC compliant.

The distance between the VFD and the motor is critical. Winding failure may occur as a result of reflected wave high voltages induced by positioning the motor far from the VFD. The logical strategy would be to keep the VFD within sight of the motor (the NEC specifies a distance of less than 50 feet and within line of sight). If this is not achievable, we can specify that the VFD have an output dv/dt filter to reduce reflections.

CHAPTER-4: EFFICIENCY OF ELECTRICAL ENERGY

HVAC systems are significant electrical energy consumers. Fans and pump motors consume more than half of the energy in an HVAC system. Through better selection and management techniques, facilities can reduce their energy expenses by 10% to 25%. Many buildings, for example, continue to utilize inefficient throttling devices such as vanes or valves to manage system flow. This is a massive waste of energy. The fan speed can be modified to vary the amount of ventilation air delivered to places with changing occupancy. A 20% reduction in the speed of centrifugal pumps and fans saves up to 50% of the energy.

This section explains the foundations of mechanical drives, which will aid in recognizing the possibilities for energy savings in HVAC systems.

PUMPS

Pumps and fans transport fluids or air from one place to another. A spinning series of impellers draws the fluid in and then pushes it out to move the fluid. The pump has a predictable performance curve for a fixed impeller diameter and speed.

Wherever the pump is functioning, the features of the system on which it is acting dictate its performance curve. The operating point is the point at which the performance curve and the system curve intersect.

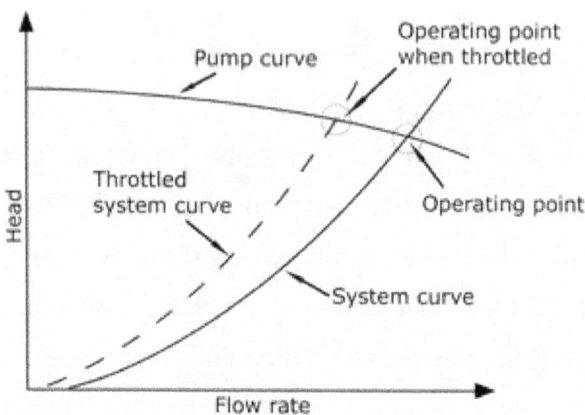

Pump and system curves cross and throttle shift the operating points

The system curve shows all of the frictional losses caused by the system's pipework, elbows, valves, and other physical components. The system curve is parabolic because its values are proportional to the flow rate squared.

A Basic Pump System:

A pump has two key characteristics. Fluid is caused to flow at a particular rate, i.e., the fluid is pushed along at a specific rate. Secondly, it will support a head of fluid.

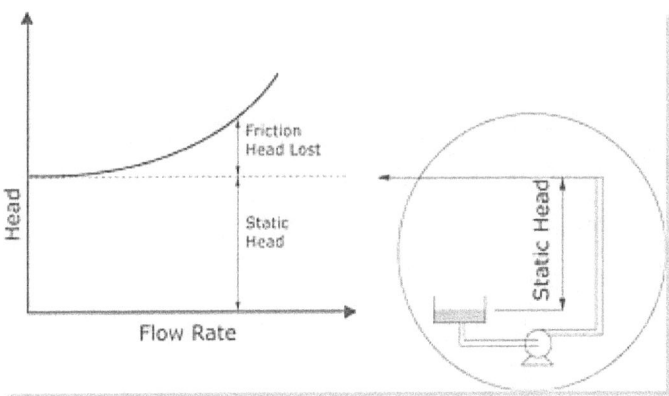

System curve showing static head and friction head loss

There are two parts as seen in the diagram:

1. Static head
2. Losses caused by friction due to the piping

As the pump curve is superimposed on the system characteristic curve, a pump is chosen for a duty. At the desired flow rate, the curves cross at the pump's best efficiency point (BEP). The crossover point is shown by the pump curve (left) and the pump curve superimposed on the system curve (right).

It is difficult and impractical to get a perfect match between the system curve and the optimal efficiency point. In reality, the pump has the potential to become unstable. Previously, the following approaches were used:

1. Adjusting the pump speed;
2. Adjusting the pump impeller diameter;
3. Changing the impeller design;
4. Adjusting the system resistance;
5. Modifying static head; and
6. Providing a system bypass flow route.

A preferred technique is to use a variable speed drive to drive the pump as this will enable the performance of the pump to be adjusted instantaneously. The VSD motor is a very efficient, low-cost method of varying the motor speed.

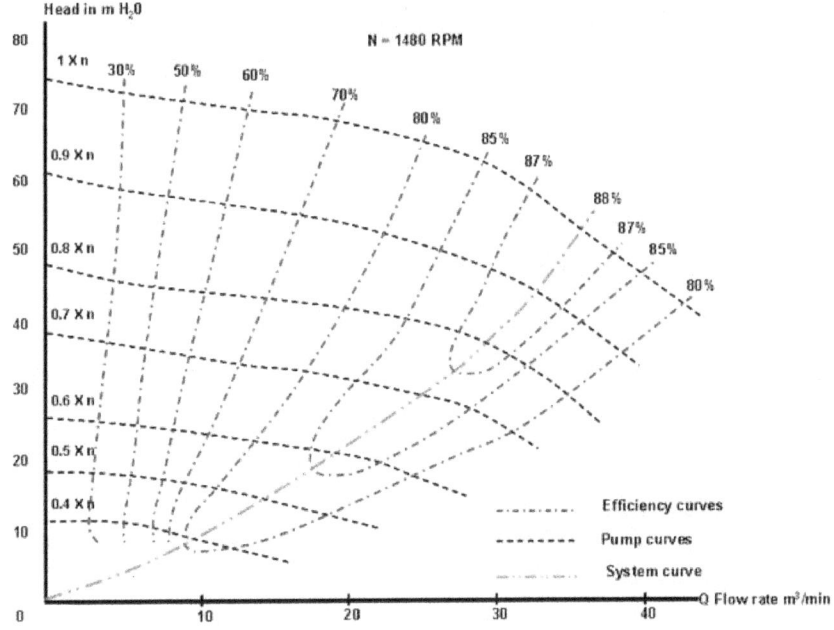

Real pump performance curves, system curves and efficiency curves

Different pumping conditions:

Two main pumping conditions are often demanded:

1. Constant pressure/flow
2. Variable flow

Let's look at each of these in turn.

Constant pressure/ flow system:

There are times when we need to keep the pressure and flow steady. For example, the water system of a hotel, the irrigation system of a farm, and the water system of a factory. The same flow is required at each outlet as new outlets are opened.

Constant flow pumping:

Flow can be controlled by using mechanical devices like throttle valves.

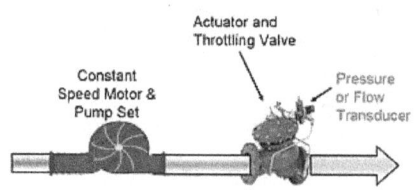

Controlling the flow rate by throttling the pump mechanism is energy inefficient. The system curve will move closer to the 'head' axis when the system is throttled. The motor continues to run at full speed even while the system is being throttled. This is where energy is being squandered. The operating point decreases as the throttle restricts flow.

Power loss from throttling:

The following graph illustrates the extent of power loss from throttling to control flow rate. Power loss is the upper grey area

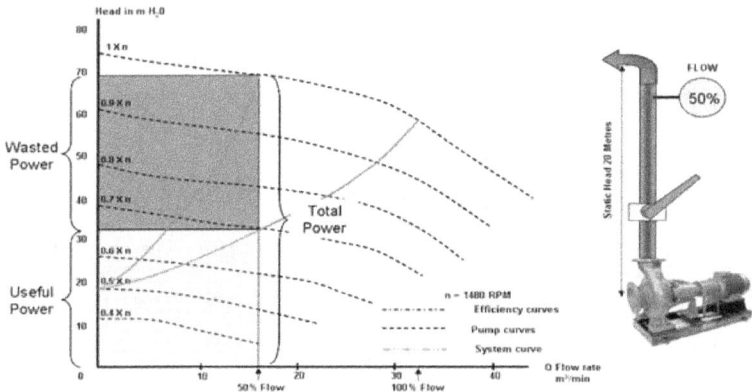

How can this energy wastage be addressed:

There are two (2) possible solutions:

1. Reduce the impeller size, or
2. Install a variable flow VSD system.

Impeller trimming to reduce power loss:

If you find a system being throttled by 12% all the time, you may reduce the impeller size and reduce your energy use by 15 to 25%. Impeller trimming resets the operating point so that flow is reduced.

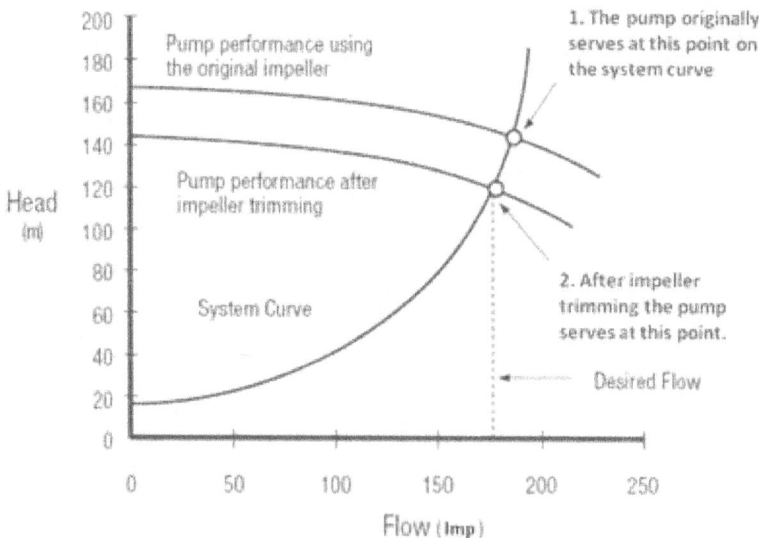

Variable flow VSD systems for better efficiency:

As the VSD system responds to changing system conditions, the motor speed is reset, i.e., as flow and pressure drop, so does the load (kW).

This ensures that the pump's operating point always follows the system curve, saving tremendous amounts of energy because only the energy needed is used. The VSD changes the operating point on the fly. This means that there is no energy loss.

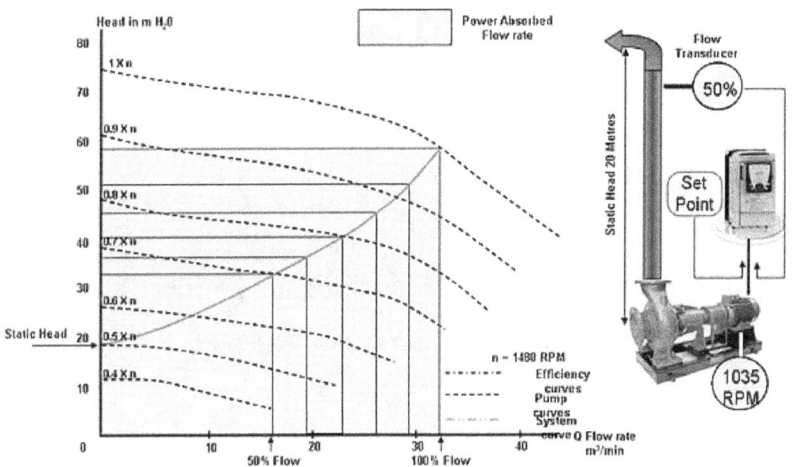

The affinity laws:

The In hydraulics and HVAC, affinity laws are used to express the relationship between numerous variables involved in pump or fan performance, such as head, volumetric flow rate, shaft speed, and power.

1. Since flow is proportional to motor speed, doubling the flow requires doubling the motor speed.
2. The pressure produced is proportional to the square of the motor speed. The pressure produced is four times more when the engine speed is doubled.
3. The necessary power is proportional to the motor speed squared. The power required to double the speed is eight times greater.

For example:

For a 100 kW pump, if the flow required is only one-half (0.5) of what is rated, then the motor could be operated at:

- half speed, and
- the pressure would become $(0.5)2 = 25\%$ of the rated flow.

The power required to operate the pump would be 100 x $(0.5)3 = 12.5$ kW.

There are several alternatives for pump control. The graph contrasts the energy consumption of each control solution. Except when near to 100% load, the three (3) curves show that VSD surpasses the use of throttles and bypass valves.

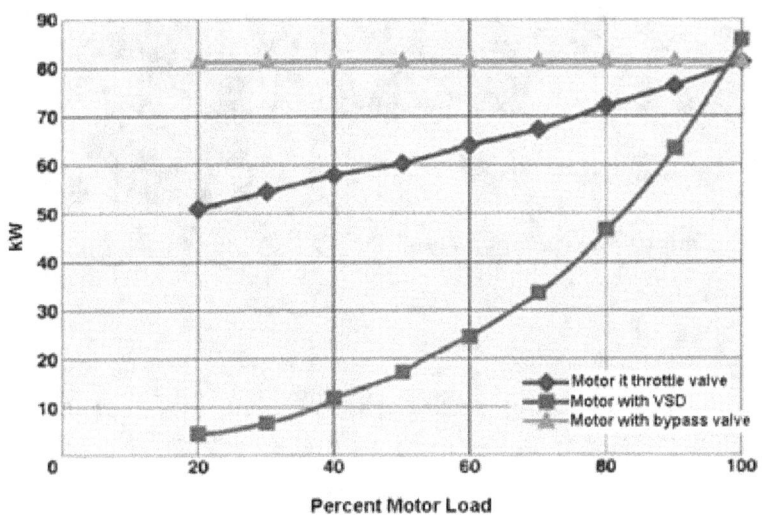

VSD power usage is lower than the mechanical flow controls

FANS

The flow of air is controlled by fans. There are several methods for controlling airflow. The figure demonstrates all three methods:

1. Guide vanes
2. Adjustable pitch blades
3. Throttling valve

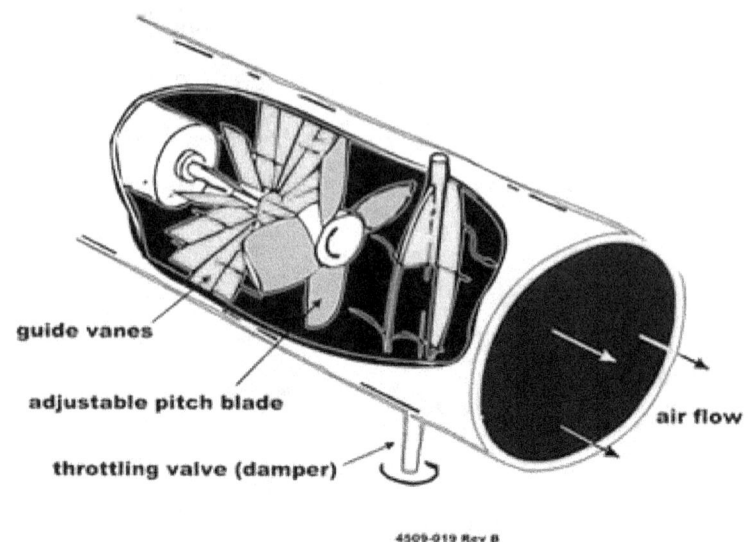

guide vanes

adjustable pitch blade

throttling valve (damper)

air flow

4509-019 Rev B

Vane, pitch blade, and throttle valve air flow control mechanisms in fans Fan performance:

A fan's performance can be graphed, as shown in the fan curve and system curve below. Not unexpectedly, the performance is comparable to that of a pump. The pressure drops as the airflow increases. The fan curve demonstrates this. The air is routed through a network of pipes and other devices to form a system. The fan's operating point is at the junction of the two curves. It should be noted that the system's frictional effect absorbs some power.

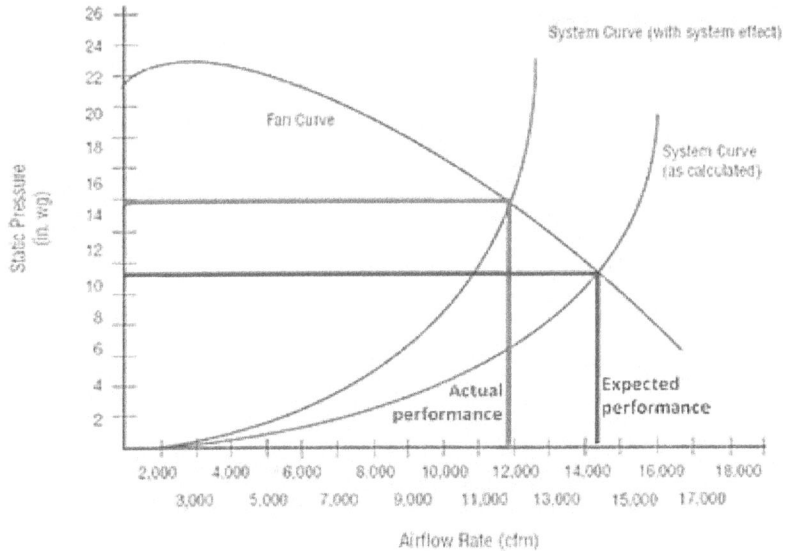

Fan curve and system curves

Power requirements for airflow control:

Each airflow control technology imposes an energy loss burden or energy losses. You may reduce energy waste by selecting the right airflow control. As seen in the picture below, the impacts change for each airflow control:

1. VSDs
2. Controllable pitch
3. Inlet vanes
4. Outlet vanes

Example calculation:

Compare the power savings of using a VSD to drive six (6) 1-kW fans. The fans feed into identical systems of duct work and fixtures. Calculate the power requirement if the airflow is reduced to 50% using either system.

Solution:

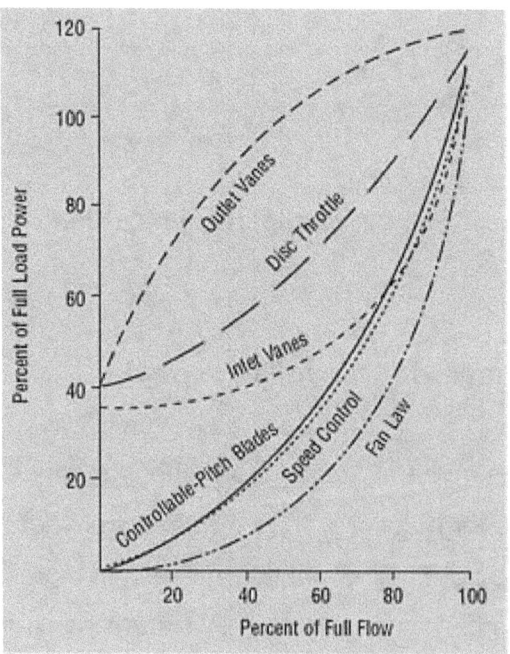

Fan and system curves to show the effects of different flow controls

To achieve half the air flow in the non VSD system simply requires the plant to switch off half of the fans, therefore power consumption is 3 x 1 kW = 3 kW.

Using the affinity law or fan law, the VSD solution can be seen to consume 0.75 kW compared to the non VSD system that consumes 3 kW.

See the diagram below:

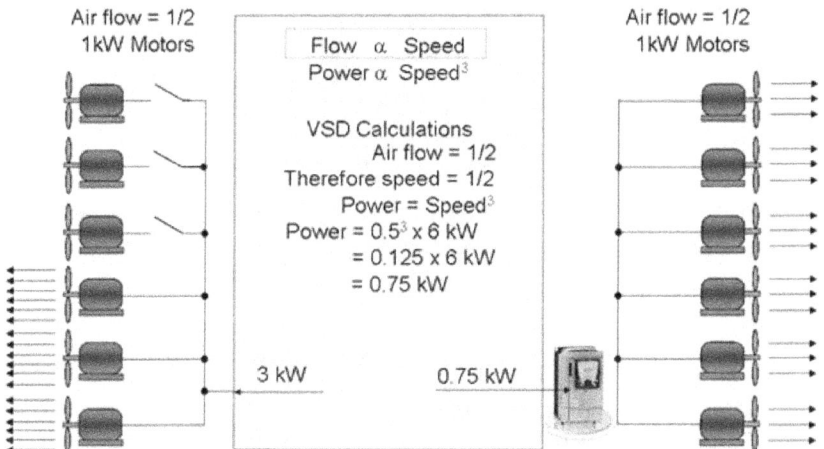

CHILLERS

Chillers often utilize the most electricity of any single appliance in a large business building. Due to lower condensing temperatures, packaged water-cooled chillers are typically 40% more efficient than air-cooled chillers in producing chilled water for use in the building's air conditioning system. Upgrading existing equipment with a variable speed drive can also drastically cut energy use in a building.

Chillers are often chosen based on their efficiency when providing 100% of their cooling capacity, but this condition is rarely met. There are several methods to express a chiller's efficiency, but the most common measure is kW of electrical input per ton (12,000 Btu/hr) of cooling produced. "kW/ton" is an abbreviation for this.

Manufacturers frequently advertise "0.55 kW/ton" chiller efficiency (or better) at full load, assuming that this translates to efficiency under all conditions. In most circumstances, knowing the efficiency throughout a range of loads from 10% to 100% is more important. The following are three strategies for increasing chiller plant load efficiency:

1. Select a chiller that can function at lower condenser water temperatures;
2. Select a variable speed drive (VSD) for the compressor motor. In practice, centrifugal chillers are only available with VSDs because they are rarely utilized with other compressor types (reciprocating, scroll, or screw compressors); and/or
3. Choose the quantity and size of chillers based on anticipated operating conditions.

How Chiller Efficiency is measured:

- Coefficient of Performance (COP) [$W_{cooling\ output}/W_{power\ input}$] - the ratio of the rate of heat removal to the rate of energy input to the compressor. Higher values correspond to improved efficiency.
- Full Load Efficiency [kW/ton] - the ratio of the rate of power input (kW) to the rate of heat removal, in tons (1 ton = 12,000 Btu/hr). Lower values correspond to improved efficiency.
- Integrated Part Load Value (IPLV) [kW/ton] - the weighted average cooling efficiency at part load capacities related to a typical season rather than a single rated condition at rating conditions specified by ARI Standard 550 or 590, depending on chiller type.
- Applied Part Load Value (APLV) [kW/ton] - calculated the same way as IPLV, but using actual chilled and condenser water temperatures rather than those specified by ARI standard rating conditions.
- Non-Standard Part Load Value (NPLV) [kW/ton] - a revision of APLV that provides a more realistic model of off-design performance.

Selecting a chiller:

When selecting a chiller, you should:

- Review your air conditioning needs: buying more capacity than you need increases your initial costs and could increase your monthly bills for the life of the equipment, which can be up to 30 years.
- Obtain competitive bids: different equipment manufacturers will offer various performances and efficiency options, many of which provide different benefits and system interactions. A consulting engineer can be employed to conduct an independent review of all selections.
- Select a chiller that operates at a lower kW//ton at part load: a chiller that is most efficient at peak load may not result in the most efficient operation over the entire cooling season. Selecting chillers that operate at a lower kW/ton at part load may be the better option.
- Expect to pay more up front: realize that this is often money well spent. Even modest improvements in chiller efficiency can yield energy savings and attractive incentives and paybacks.
- Consider selecting machines of different sizes for multiple chiller installations: select one machine small enough to meet light loads efficiently and the other(s) to meet larger loads efficiently.

COOLING TOWERS

Cooling towers are an essential component of many HVAC systems. The fans in the cooling tower remove heat from the water. To maintain the ideal condenser water temperature, the fan speed is adjusted. Since fans consume the most power in a cooling tower, they are the first area to look for energy savings.

The use of variable speed drives has the advantage of a fan's power varying as the cube of the airflow rate and the direct variation of thermal performance.

Examples of cooling towers

AIR HANDLERS

An outside unit and an indoor unit comprise the central air-conditioning system. The indoor unit is the air handler, which incorporates a coil and an air blower. Its function is to circulate the conditioned air throughout the structure. The fan is the main energy consumer in an air handling unit (AHU).

Energy savings from mechanical components:

Mechanical belts:

Flat drive belts and V-belts are less efficient because they slip more over a pulley. Furthermore, V-belt efficiency degrades with age by roughly 4%, plus further 5 to 10% if the belts are improperly maintained. V-belt oversizing and undersizing might result in additional losses. Since cogged systems have less slip, less energy is wasted in slippage. Furthermore, a cogged system requires less maintenance.

Belt alignment:

Correcting any belt misalignment will significantly reduce wear and tear on both the motor and the motor shaft. Belts should be tensioned properly. For belt drives, putting the motor on slide rails provides for easy adjustment of both alignment and belt tension. The following HVAC system diagram depicts the losses in a typical system:

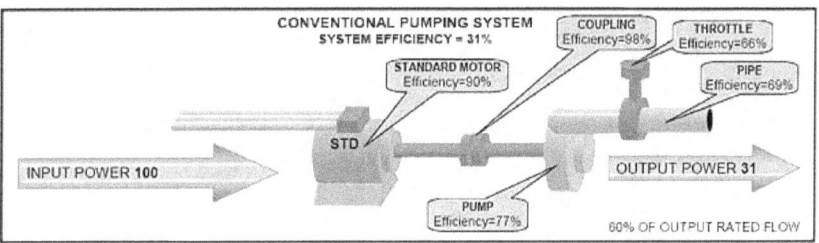

ELECTRICAL STANDARDS

There are three key organizations in the United States that serve to standardize equipment specifications and safety rules in the electrical industry:

1. The National Electric Manufacturers' Association (NEMA) is the largest, issuing technical standards and specifications that are frequently quoted in manufacturers' product descriptions.

111

2. The American National Standards Institute (ANSI) establishes utility operational standards.

3. The National Electrical Code (NEC) is a book of comprehensive building code regulations published by the National Fire Protection Association (NFPA). It is reviewed every three years, and its principal goal is to safeguard life and property.

In addition to the NEC, the NFPA publishes other electrical standards, which you should review if your region has made them law.